Plant Diseases and Pests Management

– Author –

Kalyan Jha

BIO - GREEN
BOOKS

Bio-Green Books

New Delhi - 110 002

ISBN: 978-93-90438-27-3 (Hardback)

Published by: **BIO-GREEN BOOKS**
4736/23, Ansari Road, Darya Ganj,
New Delhi – 110 002
Phone: 011-23245578, 23244987
E-mail: biogreenbooks@gmail.com

Printed at : **Replika Press Pvt. Ltd.**

Preface

Plant diseases have caused severe losses to humans in several ways. Starvation and uprooting of families resulted from the Irish famine caused by potato late blight (caused by Phytophthora infestans). A valued resource was lost with the virtual elimination of the American chestnut by chestnut blight (caused by Cryphonectria parasitica). And direct economic loss such as the estimated one billion dollars lost in one year to American corn growers from southern corn leaf blight (caused by Cochliobolus maydis, anamorph Bipolaris maydis). Many plant diseases cause less dramatic losses annually throughout the world but collectively constitute sizable losses to farmers and can reduce the aesthetic values of landscape plants and home gardens.

The goal of plant disease management is to reduce the economic and aesthetic damage caused by plant diseases. Traditionally, this has been called plant disease control, but current social and environmental values deem "control" as being absolute and the term too rigid. More multifaceted approaches to disease management, and integrated disease management, have resulted from this shift in attitude, however. Single, often severe, measures, such as pesticide applications, soil fumigation or burning are no longer in common use. Further, disease management procedures are frequently determined by disease forecasting or disease modelling rather than on either a calendar or prescription basis. Disease management might be viewed as proactive whereas disease control is reactive, although it is often difficult to distinguish between the two concepts, especially in the application of specific measures. Plant disease management practices rely on anticipating occurrence of disease and attacking vulnerable points in the disease cycle (i.e., weak links in the infection chain). Therefore, correct diagnosis of a disease is necessary to identify the pathogen, which is the real target of any disease management program. A thorough understanding of the disease cycle, including climatic and other environmental factors that influence the cycle, and cultural requirements of the host plant, are essential to effective management of any disease. The first principle includes disease management tactics applied before infection , the second principle functions with any measure applied after the plant is infected. An example of the first principle is enforcement of quarantines to prevent introduction of a disease agent into a region where it does not occur. This principle is defined as any measure that prevents the introduction of a disease-causing agent (pathogen) into a region, farm, or planting. The basic strategy assumes that most pathogens can travel only short distances without the aid of some other agent such as humans or other vector, and that natural barriers like oceans, deserts, and mountains create obstacles to their natural spread. In many cases pathogens are moved with their host plants or even on nonhost material such as soil, packing material or shipping containers. Unfortunately, exclusion measures usually only delay the entry of a pathogen, although exclusion may provide time to plan how to manage the pathogen when it ultimately arrives. Karnal bunt of wheat is an example of a pathogen originally from India that was anticipated. Measures were established to prevent its introduction, but it finally found its way into the United States. Soybean rust has been found recently in the southeastern U.S. and precautions have been undertaken to prevent further spread. Due to its destructiveness, South American leaf blight is a feared disease in the major rubber producing region of Indonesia, and contingency plans have been proposed to chemically defoliate rubber trees by aerial application of herbicides if the pathogen is detected. It is hoped that this would prevent establishment of the pathogen in the region. An important and practical strategy for excluding pathogens is to produce pathogen-free seed or planting stock through certification programs for seeds and vegetatively propagated plant materials such as potatoes, grapes, tree fruits, etc. These programs utilize technologies that include isolation of production areas, field inspections, and removal of suspect plants to produce and maintain pathogen-free stocks. Planting stock that is freed of pathogens can be increased by tissue culture and micropropagation techniques as well as be maintained in protective enclosures such as screenhouses to exclude pathogens and their vectors.

In a nutshell, this book covers principles involved in the control of plant diseases and points to research avenues that might lead to improved methods for an environmentally safe control approach in the near future. This volume will definitely serve not only as an excellent reference material but also as a practical guide for teachers, research workers, students and plant pathologists as well.

—Dr. Kalyan Jha

Contents

Chapter

1

Infectious Plant Diseases and their Control

A systems approach to plant health includes preventing and controlling diseases. There is one thing that must be kept in mind to be successful at plant disease control. You do not control a particular disease by doing any one particular thing. Diseases are managed, prevented or controlled by groups of practices loosely grouped under the concepts of the systems approach.

Infectious organisms, called plant pathogens, are part of the environment whenever and wherever plants are in existence. Surely, we do try and practice sanitation to limit their numbers. However, we cannot achieve total "sterility" of our greenhouses. We must manage the environments of our crops so plant pathogens cannot develop and flourish.

Cultivated plants are usually more susceptible to disease than their wild relatives, partly because large numbers of the same plant are often grown closely together in pure stands. Disease-causing organisms (pathogens) often get established under these conditions. Once this happens, they may spread rapidly. In addition, many of our valuable crop and ornamental plants are basically very susceptible to disease and would have difficulty surviving in undisturbed nature. Finally, cultivation – be it geraniums, cabbages or oak trees – constantly disturbs nature and tends to create environmental stresses. Stresses often weaken plants and subject them to pathogenic infections.

What Causes a Plant Disease?

A plant can become diseased when it is continuously disturbed over a fairly long period of time by some factor or group of factors in its environment. Sometimes one of these disturbing factors is a living, infectious pathogen. Living pathogens include bacteria, fungi, viruses, nematodes and parasitic seed plants. Only if the disturbances are severe enough to produce noticed effects, called symptoms of poor health, do we call the plant diseased. Obviously, knowledge of normal growth habits, variety (cultivar) characteristics and normal variability of plants within a species is required for recognition of a diseased condition.

Figure: *Protect cuttings from drying out before sticking to help control stem rots.*

A System of Plant Disease Management

Employing health management systems is the most important way to proceed and succeed in growing plants profitably in a greenhouse or nursery. These are not complicated or difficult. They are simply based on the fact that a plant should not be chronically disturbed by elements of its environment with which it is has difficulty coping. Such plants are stressed and are more subject to the infection and development of infectious pathogens. Reduction of plant stress involves awareness of environmental disturbances. It also involves trying to set up and maintain a reasonably balanced environment for each and every type of plant in the greenhouse or nursery.

Beyond Stress Management

As you have no doubt realized, when you are growing dozens of different kinds of plants, it is impossible to grow each one at its precisely balanced environment. Furthermore, there are some things you cannot control as you would like.

The amount of sunlight is probably the environmental factor that gives you the most trouble. Other things happen as a result of new workers tending to the plants differently, equipment breakdowns or changes in fertilizers, growing media, etc. In other words, a bumper sticker saying "Stress Happens" might be a good one for greenhouse growers!

Pathogen Management Practices

When stress happens, plant pathogens are primed to strike. We must turn our attention to methods of dealing with pathogens. Successful pathogen management is based on accurate diagnosis of the cause, thorough knowledge of the pathogen and its disease cycle, how the host and pathogen interact with various environmental factors, practicality of possible actions and cost of various practices. The most important point in controlling a plant disease is choosing the best methods for a given situation. The best methods of control for one type of disease on a certain host may not be the best methods for another disease on the same or a different plant. Also, the use of several control measures (integrated control) is often needed. For example, several cultural control practices are often combined with a protective fungicide spray, sprench or drench program.

Infectious diseases in greenhouses or nurseries are controlled by one of four basic methods:

Exclusion. Disease-causing organisms should be excluded from production areas whenever possible. Any plant pathologist will have the practice of purchasing only pathogen-free plants, cuttings or seed high on his or her list. So what else is new? The problem is that you cannot be certain you have done this, even if you do pay extra money for "indexed" or "cultured" material. Try to do the best you can, of course. Sanitation in and around the greenhouse is an excellent pathogen exclusion measure, as well. It involves cleaning and disinfecting potting benches, soil bins, head houses, greenhouse benches, tools and equipment. Disinfecting tools and equipment is a most valuable practice to help prevent the spread of many viruses and bacterial diseases. Growing crops on raised benches is an important way to keep pathogens away from crops. Using potting media that is relatively free of plant pathogens is yet another exclusion practice. Controlling weeds that harbour insects to spread pathogens or offer survival for pathogens between crops is important.

There are other exclusion practices that serve to prevent spread of pathogens from one area to another in the greenhouse or on any particular bed or bench. The most important of these is to prevent a great number of persons from handling stock plants. Another is to avoid splashing water from plant to plant whenever possible. Protection. Growing resistant varieties is a form of protection. The genetics of the "host" protect it from pathogen infection or development. This is the most common form of protection from disease. After all, most plants do not get most diseases.

Most of us readily recognize that plants can be protected from pathogens by uniform and timely applications of disease-control chemicals (fungicides, bactericides and nematicides). Diseases caused by fungi are most commonly treated in this manner. Whereas there are many specific products that are particularly useful for particular fungal diseases, the trend of late is to use broad-spectrum products or combinations of products for general fungal disease treatment when necessary.

For leaf and flower diseases, a combination of mancozeb or chlorothalonil (contact protectants) and thiophanate-methyl (a systemic protectant) is very popular now. The new strobilurins are very useful. Flutolanil is coming into widespread use now. The group of DMI systemics are useful for powdery mildews, rusts and other diseases. For root rots, a combination of thiophanate-methyl, fludioxinal or iprodione and etridiazole or mefenoxam is commonly used. When applying fungicides, it is important to have the fungicide on the plant surface or in the soil before infection takes place. Most fungicides are protectants, not eradicants.

Prevention of pathogen infection and development conditions. Many of these practices come under the stress management practices discussed earlier. It never ceases to amaze me how tough plants are if they are not stressed. Many commonly suggested cultural practices serve to manage infectious plant diseases through the idea of preventing spread, infection and development of pathogens. These generally involve altering the air and soil environment. Keeping plants spaced or growing plants on heated floors to allow foliage to dry quickly after watering are important examples. Spacing plants ensures adequate light to counter stress. It also favors uptake of needed nutrients for good plant health.

For root health, avoid overwatering and persistent wetness of the growing media. Also, avoid excessive dryness between waterings, use well-aerated and well-drained growing media and do not allow soluble salts to build up. Injured roots favour root-rotting pathogens.

Eradication. Plant pathogens can be eliminated (eradicated) and insects that spread pathogens can be controlled by pesticides. Removal and destruction of diseased plants is a common way to help control systemic bacterial and viral diseases. These plants may be the crop you are growing, weeds or other, less important host plants. Other eradication methods include crop rotation, which "starves" out soil-invading pathogens in nurseries. As you can see from the above remarks, most of the disease-controlling systems we use today do not involve spraying or drenching with chemicals. These growing practices are effective and widely used by good growers. They take planning and follow through. This is truly a systems approach to plant health management.

Bacterial Diseases

Colibacillosis (Coliform Infections)

Problems attributed to coliform infections are often caused by strains of the *Escherichia coli* organism. There is a marked variation in severity. Problems range from severe acute infections with sudden and high mortality to mild infections of a chronic nature with low morbidity and mortality. Infections may result in a respiratory disease from air sac infection, a septicemic (blood) disease from generalized infections, an enteritis from intestinal infection or a combination of any or all of these conditions. The disease may result from a coliform infection alone as in primary infection or in combination with other disease agents as a complicating or secondary infection. Secondary infections commonly occur as a part of the classic air sac disease syndrome as a complication with *Mycoplasma gallisepticum* infections.

All ages can be affected; however, the acute septicemia in young turkeys and airsacculitis in young chickens is more common in young growing birds. High, early mortality may occur as the result of navel infections. The symptoms of this disease is caused by the *E. coli* bacteria and the toxins produced as they grow and multiply. There are many different strains or serological types within the group of *E. coli* bacteria. Many are normal inhabitants in intestinal tracts of chickens and turkeys and consequently are common organisms in the birds' environment.

A marked variation exists between different strains in their ability to cause disease. Some are severe and by themselves can cause disease while others are supposedly harmless. All degrees of pathogenicity exist between the two extremes.

The primary routes of invasion by the organism are the respiratory system and the gastrointestinal tract. Omphalitis and infections in young birds may result from entry through the unhealed navel or penetration of the egg shell prior to or during incubation. The symptoms vary with the different types of infections. In the acute septicemic form, mortality may begin suddenly and progress rapidly. Morbidity may not be apparent and birds in apparently good condition may die. However, in most cases birds are listless with ruffled feathers and indications of fever. Additional symptoms of laboured breathing, occasional coughing and rales may be apparent. Diarrhea may be evident. Mortality may be high in recently hatched chicks and poults as a result of navel infection of coliforms.

Extremely acute septicemic infection may result in sudden death with very few, if any, lesions apparent. Common lesions include dehydration, swelling and congestion of the liver and spleen and kidneys and pinpoint hemorrhages in the viscera. Fibrinous to caseous exudate in the air sacs, heart sac and on the surface of the heart, liver and lungs is a characteristic lesion. The intestines may be thickened and inflamed and may contain excess mucus and areas of hemorrhage. Navel infections, similar to those described for omphalitis may be seen in young birds. Diagnosis by laboratory means is necessary since coliform infection in its various forms may resemble and be easily confused with many other diseases. Isolation and identification of the organism by culture procedures can be accomplished relatively quickly; however, mere isolation is not sufficient to make a diagnosis. One must take into consideration the organ from which the organisms were isolated, the pathogenicity of the particular isolate and the presence of other disease agents.

Management and sanitation practices designed to reduce the number of these types of organisms in the birds' environment are necessary. In addition, reducing stress factors and other disease agents can enhance the ability of birds to defend against harmful infections. Providing adequate ventilation, good litter and range conditions, properly cleaned and disinfected equipment and facilities and high quality feed and water will improve the disease resisting status of the birds. The poultryman must always avoid overcrowding, environmental stresses like chilling or overheating and avoid vaccination or handling stress during periods when the birds are already subjected to stressful conditions. Proper egg handling, good hatchery management and implementing a good sanitation program is necessary to reduce early exposure of chicks or poults to disease organisms. It is always emphasized that problems due to one of the more pathogenic strains may occur even under ideal conditions.

The response of coliform infections to various medications is erratic and often difficult to evaluate. Under practical conditions, treatment is often disappointing. Drug sensitivity varies with the strain of *E. coli* causing the condition. Laboratory tests to determine the sensitivity to the various drugs are useful to select the most beneficial drugs. When practical, moving birds to a clean environment may be of more value than medication. For example, when outbreaks occur in growing turkeys in the brooder house, moving to range is often the best treatment.

Mycoplasmosis (CRD, Air sac, Sinusitis)

Organisms in the genus Mycoplasma are a significant cause of respiratory disease in poultry. Of the numerous species of Mycoplasma that have been isolated from domestic poultry, three are of known significance. *Mycoplasma gallisepticum* is associated with chronic respiratory disease (CRD)/air sac syndrome in chickens and turkeys and infectious sinusitis of turkeys; *Mycoplasma meleagridis* is associated with airsacculitis in turkeys; and *Mycoplasma synoviae* is the cause of infectious synovitis in chickens and turkeys.

Chronic respiratory disease (CRD), air sac syndrome and infectious sinusitis of turkeys have a common cause. CRD was first recognized as a chronic but mild respiratory disease of adult chickens. It reduced egg production but caused little or no mortality. Afterward, a condition known as "air sac disease" became a problem in young birds. It caused high mortality in some flocks. Many birds became stunted, feed efficiency was reduced, and many fowl were rejected as unfit for human consumption when processed. Infectious sinusitis in turkeys produces a sinus swelling under the eye as well as an inflammation of respiratory organs. It is a chronic disease adversely affecting growth and feed conversion. It may also cause significant mortality in young poults.

A peculiar bacterial-like organism known as *Mycoplasma gallisepticum* (MG) is common to all three conditions. CRD and sinusitis in turkeys are caused by a pure MG infections while the air sac syndrome is caused by an infection of MG in combination with *E. coli*. These conditions are triggered by acute respiratory infections such as Newcastle disease or infectious bronchitis.

Mycoplasma gallisepticum is widespread and affects many species of birds. Eradication programs have reduced the incidence in recent years. It is primarily spread through the egg. Infected hens transmit organisms and the chick or poult is infected when it hatches. Organisms may also be transmitted by direct contact with infected or carrier birds. The true CRD produces slight respiratory symptoms such as coughing, sneezing and a nasal discharge. In the air sac syndrome there is an extensive involvement of the entire respiratory system. The air sacs are often cloudy and contain large amounts of exudate. Affected birds become droopy, feed consumption decreases and there is a rapid loss of body weight.

Infectious sinusitis in turkeys occurs in two forms. When the "upper" form is present, there is only a swelling of the sinus under the eye. In the "lower" form, the lungs and air sacs are involved. The air sacs become cloudy and may contain large amounts of exudate. Both forms of the disease are usually present in the flock and frequently are present in the same bird. Diagnosis of either condition must be based on flock history, symptoms and lesions. Blood tests are useful in determining whether a flock is infected.

The answer to the MG problem in both chickens and turkeys is eradication of the disease organisms. This goal has been achieved in commercial breeding flocks with voluntary programs conducted by the National Poultry Improvement Plan (NPIP) and National Turkey Improvement Plan (NTIP). The treatment of CRD, air sac syndrome and the lower form of infectious sinusitis is not considered satisfactory. Many antibiotics have been used with varying success. Whether to give treatment is a decision that must be made on each flock based on economic factors. If treatment is attempted, give high levels of one of the broad spectrum antibiotics (Tylosin, aureomycin, terramycin, gallimycin) either in feed, drinking water or by injections. The "upper"; form of infectious sinusitis can be treated with success by injecting antibiotics into the swollen sinus cavity.

Fowl Cholera

This disease occurs throughout the country wherever poultry is produced and in recent years has become the most hazardous infectious disease of turkeys. Host range is extensive and includes chickens, turkeys, pheasants, pigeons, waterfowl, sparrows and other free-flying birds. The causative organism of fowl cholera is *Pasteurella multocida*. The organism can survive at least one month in droppings, three months in decaying carcasses and two to three months in soil. Pasteurella apparently enters tissues of the mouth and upper respiratory tract. The disease is not transmitted through the egg.

Major sources of infection include:

- Body excreta of diseased birds that contaminate soil, water, feed, etc.,
- Carcasses of birds that have died of the disease,
- Contaminated water supplies such as surface tanks, ponds, lakes and streams,
- Mechanical transmission by contaminated shoes or equipment.

Studies indicate that animals other than birds may serve as reservoirs of infection and actively spread the disease. These animals include raccoons, opossums, dogs, cats, pigs, and vermin.

The disease is seldom seen in chickens under four months of age, but is commonly seen in turkeys under this age. In the peracute form, symptoms may be absent; in the acute form some birds may die without showing symptoms, but many others are visibly ill before death. Characteristic symptoms include stupor, loss of appetite, rapid weight loss, lameness resulting from joint infection, swollen wattles, difficult breathing, watery yellowish or green diarrhea and cyanosis or darkening of the head and wattles.

Lesions may be lacking in birds dying during peracute outbreaks. When present, lesions may resemble those associated with any acute septicemic bacterial infection, often those of fowl typhoid. Typical lesions may include pinpoint hemorrhages in the mucous and serous membranes and/or abdominal fat; inflammation of the upper third of the small intestine; light, firm "parboiled" appearance of the liver; enlarged and congested spleen; creamy or solid collection of material in joints; and cheesy material in the internal ear and air spaces of the cranium of birds having twisted necks. Turkeys may have pneumonia with solidification of one or both lungs. A tentative diagnosis may be made on flock history, symptoms and postmortem lesions. A definite diagnosis depends upon isolation and identification of the organism.

Properly administered bacterins are helpful in preventing fowl cholera, particularly in turkeys. Their use must be combined with a rigid program of sanitation. In general, as it applies to the use of bacterins in turkeys, complete protection is unrealistic. Follow the manufacturer's recommendations for use of the bacterin. Vaccination in conjunction with treatment is not recommended. Sanitation practices that aid in preventing the disease are:

- Complete depopulation each year with definite breaks between older birds and their replacements,

- Implement a rodent control program,
- Dispose of dead birds properly,
- Provide safe, sanitary water,
- Clean and disinfect all houses and equipment after disposing of flock,
- Keep birds confined to the house and away from wild feral birds and animals,
- Allow contaminated ranges or yards to remain vacant for at least three months.

Although drugs usually alter the course of a fowl cholera outbreak, affected birds remain carriers and the disease has a tendency to recur when treatment is discontinued. This may necessitate prolonged treatment with drugs added to the feed and water. Sulfa drugs and broad spectrum antibiotics (Penicillin) usually control losses.

Necrotic Enteritis

Necrotic enteritis is an acute disease that produces a marked destruction of the intestinal lining of the digestive tract. Common field names (rot gut, crud and cauliflower gut) accurately describe the condition. The cause of the disease is *Clostridium perfringens*, a spore-forming, rod-shaped bacterium. Bacterial organisms and their toxins are the primary cause but coccidiosis may be a contributing factor. Most of the damage to the intestinal lining apparently is due to toxins produced by the bacterial organisms. Little is known about the spread of the disease but transmission is thought to occur by oral contact with the droppings from infected birds. Necrotic enteritis appears suddenly in the affected flock. Apparently healthy birds may become acutely depressed and die within hours. Mortality is usually between two and ten percent, but may be as high as thirty percent in severe outbreaks. Losses due to reduced growth and feed conversion may be more costly than flock mortality.

Lesions of the disease usually involve the lower half of the small intestine, but in some instances the entire length of the tract is involved. The intestine is dilated, contains dark offensive fluid and a diphtheritic cauliflower-like membrane that involves the mucosa. The lining of the intestine will have a coarse Turkish-towel appearance and portions of the lining may slough off and pass out with the intestinal contents. Diagnosis in based upon history, symptoms and findings of the characteristic lesions.

Bacitracin or virginiamycin are effective treatments administered in the feed. Bacitracin can also be given in the drinking water. Supportive vitamin treatment may enhance the effectiveness of the treatments. Preventive medication may be of value on premises where prior infections have been observed. Since coccidiosis may be a contributing factor, attention must be given to an effective coccidiosis control program.

Ulcerative Enteritis (Quail Disease)

Ulcerative enteritis is an acute or chronic infection of game birds, chickens, turkeys and other domestic fowl. Death losses may be high for young quail or pullets being raised for egg production. The cause of the disease is *Clostridium colinum*, a spore forming bacterial rod. The infection spreads by the droppings from sick or carrier birds to healthy birds. The disease organism is very resistant to disinfectants and will persist under varying environmental conditions. Birds with the acute form may die suddenly while in good flesh, whereas more chronically affected birds become listless, have ruffled feathers, whitish watery diarrhea, and develop a humped-up posture. Such birds usually die in an extremely emaciated condition. The dropping may be confused with those of birds with coccidiosis and the two diseases are often seen in the same bird. Droppings of birds with only ulcerative enteritis never contain blood.

The postmortem lesions are characteristic. The entire intestinal tract often has button-like ulcers but the lower portion is most often affected. These ulcers often perforate, resulting in local or generalized peritonitis.

Although the disease is characteristic in nature, anyone suspecting the infection should seek professional confirmation before treatment is started. Bacitracin and penicillin are the most effective drugs in the treatment and prevention of this disease. If bacitracin is used, it should be incorporated in the feed at levels up to 200 grams per ton of feed. Addition of bacitracin to the water at the rate of one teaspoon per gallon aids in controlling an outbreak of the disease. Either method of administering bacitracin will control the disease within two weeks, unless a bacitracin-resistant strain of the disease organism is present. Penicillin is also used to treat the disease if bacitracin is not effective. Raising birds on wire is an effective preventative measure. Specific drugs (bacitracin or penicillin) fed at low levels, are effective for controlling the disease in operations where the use of wire flooring is impractical.

Pullorum Disease

Pullorum disease is an acute or chronic infectious, bacterial disease affecting primarily chickens and turkeys, but most domestic and wild fowl can be infected. The cause is a bacterium named *Salmonella pullorum*. This organism is primarily egg transmitted, but transmission may occur by other means such as:

- Infected hen to egg, egg to chick, or chick to chick in incubator, chick box, brooder, or house. Survivors become infected breeders (cycle begins again),
- Mechanical transmission (carried around on clothes, shoes or equipment),
- Carrier birds (apparently healthy birds shed the disease organisms),
- Contaminated premises (from previous outbreaks).

Disease organisms may enter the bird through the respiratory (as in the incubator) or digestive systems. Most outbreaks of acute pullorum disease in chickens or turkeys result from infection while in the hatchery. Pullorum disease is highly fatal to young chicks or poults, but mature birds are more resistant. Young birds may die soon after hatching without exhibiting any observable signs. Most acute outbreaks occur in birds that are under three weeks of age. Mortality in such outbreaks may approach ninety percent if untreated. Survivors are usually stunted and unthrifty. Infection in young birds may be indicated by droopiness, ruffled feathers, a chilled appearance with birds huddling near a source of heat, laboured breathing, and presence of a white diarrhea with a "pasted-down" appearance around the vent. The white diarrhea symptom instigated the term "bacillary white diarrhea" that was commonly associated with this disease at one time. Gross lesions may be lacking in some adult birds.

Diagnosis in young birds is made by isolating the causative organism in the laboratory. In older birds, blood testing may indicate an infection but a positive diagnosis depends upon isolation and identification of the organism by laboratory procedures. Complete eradication is the only sound way to prevent pullorum disease. All hatchery supply flocks should be tested and only pullorum-free flocks used as a source of hatching eggs. Purchase chicks or poults from hatcheries that are officially recognized as "Pullorum Clean" by National Poultry Improvement Plan representatives in your state.

Treatment primarily is a salvage operation and does not prevent birds from becoming carriers. Consequently, do not keep recovered flocks for egg production. Among the drugs used to treat pullorum disease are furazolidone, gentamycin sulfate, and sulfa drugs (sulfadimethoxine, sulfamethazine, and sulfamerazine).

Fowl Typhoid

Fowl typhoid is an infectious, contagious bacterial disease that is usually acute but sometimes chronic. It affects most domestic and wild fowl including chickens, turkeys, ducks, pigeons, pheasants and other game birds. It must not be confused with typhoid fever in humans that is caused by a distinctly different organism. The cause in the bacterium, *Salmonella gallinarum*. Methods of transmission are the same as for pullorum disease, including egg transmission. However, mechanical transmission is more prevalent with this disease than with pullorum disease. Any age bird can be infected, but the disease primarily occurs in young adults (usually those older than twelve weeks of age). Mortality varies from less than one to about forty percent, but higher mortality has been observed. Signs include sudden or sporadic mortality, listlessness, green or yellow diarrhea (accompanied with pasting of the vent feathers), loss of appetite, increased thirst and a pale, anaemic appearance of comb and wattles.

A tentative diagnosis may be made from the history, signs and lesions but final diagnosis must be based on isolation and identification of the causative organism. Lesions observed at necropsy help verify a diagnosis of fowl typhoid. The lesions include an enlarged and mottled spleen, enlarged liver (coloured yellow or greenish brown), small pinpoint hemorrhages in muscles and fat surrounding internal organs, and a slimy inflammation of the front third of the small intestine. Small, white plaque-like areas are visible through the walls of the intestine is suggestive of fowl typhoid in turkeys. Blood tests used to detect pullorum reactors are also used to identify fowl typhoid birds. Prevention and control depend heavily upon basic disease prevention practices including the hatching chicks from disease-free flocks (as determined by pullorum testing), practicing strict sanitation on the farm, providing clean feed and water, and proper disposal of all dead birds as approved by the state animal health agency. The causative organism can live outside the bird body for at least six months, thus requiring extra management precautions to break the disease cycle. Following an outbreak, thoroughly clean and disinfect the premises. When feasible, practice range rotation and other special precautions to prevent the carryover of infection to the following flock.

Drugs cannot be depended upon as a means of typhoid prevention and are not recommended for that purpose. Infected birds may be salvaged using the same drugs as used to salvage pullorum infected birds.

Botulism

Botulism is a disease caused by the ingestion of a toxin produced by the *Clostridium botulinum* bacterium. All domestic fowl and most wild birds are susceptible to the toxin's effects. Many human deaths have also been attributed to the consumption of food or water containing the toxin.

Botulism is not a bacterial infection, but a condition produced by a byproduct of the bacteria's growth. The organism is common in nature and is widely dispersed in soils. Ingestion of the organism is not harmful. It becomes dangerous only when conditions are favorable for its growth and subsequent toxin formation. The organism grows best under high humidity and relatively high temperature and in an environment containing decaying organic material (plant or animal). The organism requires an environment in which all atmospheric oxygen is eliminated. The organism cannot multiply in the presence of air. Stagnant pools or damp areas with buried decaying matter are danger areas for toxin development. Botulism results after the decaying animal or plant material containing the toxin is consumed. Decaying carcasses are a frequent source of the toxin, as are many insects feeding in the same tissue. The insects may contain enough toxin to cause the disease in any bird that ingests it. Since the toxin is water soluble, water sources may become contaminated and provide a reservoir for the disease.

The toxin is one of the most potent discovered by scientists. The toxin is relatively heat stable but may be destroyed by boiling. There are different types of the toxin; types A and C cause the disease in birds while type B frequently produces the disease in man.

Weakness is generally the first sign of the illness and is followed by progressive flaccid paralysis of the legs, wings and neck. When neck muscles are affected the head hangs limp, thus causing a condition referred to as "limberneck". Affected birds may have a peculiar trembling, loose feathers that are pulled out easily and dull partly closed eyes. Some birds (turkey) do not develop loose feathers or limberneck symptoms. Because of the paralysis, birds are unable to swallow and mucous accumulates in the mouth. Fatally affected birds may lie in a profound coma appearing lifeless for several hours before death. Significant lesions are not usually observed in affected birds. Examining digestive contents may reveal insects, decomposed animal or vegetable material or other matter suggesting that the birds have consumed the toxin. A tentative diagnosis can be made from the history, symptoms and post-mortem findings. As an aid to diagnosis, sick birds may be given water into the crop, kept in a cool environment and treated intravenously with antitoxin. Recovery of a large percentage of the affected birds would substantiate diagnosis.

Prevention should be aimed at eliminating sources of toxin production and preventing access of birds to such materials. These practices include prompt removal of all dead animals from houses and pens, debeaking the birds, controlling fly and insect populations and avoiding access to decaying organic material. Contaminated water supplies are particularly dangerous.

If the disease strikes, locate and remove the source of the toxin and separate all visibly affected birds from the flock for treatment. Place sick birds in a cool shaded area and give fresh water into the crop, twice daily. Mild laxatives may be used for birds that have been exposed but do not show disease symptoms. Epsom salts (one pound per 100 birds) may be mixed into feed. Adding a level teaspoonful of Epsom salts in one ounce of water and placing in the crops of sick birds has been beneficial in many instances. Antitoxin therapy is indicated only in birds that have high individual value since the antitoxin is difficult to obtain and is expensive.

Infectious Coryza

Infectious coryza is a specific respiratory disease in chickens that occurs most often in semi-mature or adult birds. Infection may result in a slow-spreading, chronic disease that affects only a small number of birds at one time, or in a rapid spreading disease with a higher percentage of birds being affected. The occurrence of infectious coryza is not widespread and the incidence is relatively low.

The disease is caused by a bacterium known as *Hemophilus gallinarum*. Outbreaks usually result from the introduction of infected or carrier birds into a flock. Transmission of the infection occurs by direct contact, airborne infection by dust or respiratory discharge droplets and drinking water contaminated by infective nasal exudate. Susceptible birds usually develop symptoms within three days after exposure to the disease. Recovered individuals may appear normal but remain carriers of the organism for long periods. Once a flock is infected, all birds must be considered as carriers.

The most characteristic symptoms of infectious coryza include edematous swelling of the face around the eyes and wattles, nasal discharge and swollen sinuses. Watery discharge from the eyes frequently results in the lids adhering together. Vision may be affected because of the swelling. The disease results in a decrease in feed and water consumption and an increase in the number of cull birds. An adverse effect on egg production usually occurs in proportion to the number of affected birds. Diagnosis can be confirmed only by isolation and identification of the causative organism. The organism, *Hemophilus gallinarum*, is extremely fastidious and often difficult to isolate.

Prevention is the only sound approach in controlling infectious coryza. It usually can be prevented by management programs that eliminate contact between susceptible and infected birds. It requires only separating affected or carrier birds from the susceptible population. In order to prevent the infection, introduce started or adult birds only from sources known to be free of the infection. If infection occurs, complete depopulation followed by thorough cleaning/disinfecting is the only means for eliminating the disease. A number of drugs are effective for treating the symptoms of the disease although the disease is never completely eliminated. Sulfadimethoxine or sulfathiazole in the feed or water or erythromycin administered in the drinking water can reduce the symptoms of this disease.

Omphalitis

Omphalitis may be defined technically as an inflammation of the navel. As commonly used, the term refers to improper closure of the navel with subsequent bacterial infection (navel ill; mushy chick disease). Apparently, most problems result from mixed bacterial infections including the common coliforms and various species belonging to the genera Staphylococcus, Streptococcus, Proteus, and others. Omphalitis can usually be traced to faulty incubation, poor hatchery sanitation or chilling/overheating soon after hatching (such as in transit). The significance of isolating one of the bacterial species mentioned above is complicated in that many of the same species can be isolated from the yolks of supposedly normal birds immediately after hatching.

Omphalitis occurs during the first few days of life, so it cannot be considered transmissible from bird to bird. It is transmitted from unsanitary equipment in the hatchery to newly hatched birds having unhealed navels. Affected chicks usually appear drowsy or droopy with the down being "puffed up". They also generally appear to be of inferior quality and show a lack of uniformity. Many individuals stand near the heat source and are indifferent to feed or water.

Diarrhea sometimes occurs. Mortality usually begins within 24 hours and peaks by five to seven days. Characteristic lesions are poorly healed navels, subcutaneous edema, bluish colour of the abdominal muscles around the navel and unabsorbed yolk material that often has a putrid odor. Often yolks are ruptured and peritonitis is common. A tentative diagnosis can be made on the basis of history and lesions. The presence of mixed bacterial infections and absence of any specific disease-producing agent is used for confirming the diagnosis. Good management and sanitation procedures in the hatchery and during the first few days following hatching are the only sure ways to prevent omphalitis. Broad spectrum antibiotics help reduce mortality and stunting in affected groups, but they do not replace sanitation.

Erysipelas

Erysipelas is a bacterial disease caused by *Erysipelothrix insidiosa*. The disease affects several species of birds including chickens, ducks and geese, but the fowl in which it has been of primary importance is the turkey. Man is susceptible to infection and may contract the disease from infected turkeys. Since this organism is pathogenic for man, care should be taken when handling infected birds or tissues. Erysipelas in turkeys occurs most often during the fall and winter months and usually affects birds that are four to seven months of age, although any age bird is susceptible. Incidence has often been reported to be higher in males than in females, possibly because fighting males receive numerous skin abrasions that serve as portals of entry for the bacteria. In some instances the incidence is higher in hens than toms because of artificial insemination techniques that provide a means of transmission.

The organism may survive for long periods in the soil and most outbreaks are thought to originate from contaminated soil or premises. Sheep, swine and rodents may be carriers of the disease organisms. Recurrence of the disease on a premise is common. Predisposing or aggravating factors include over-crowding damp or inclement weather and poor sanitation and range management. The first indication of the disease may be the discovery of several dead birds. Usually several morbid birds can be found; however, most affected birds are visibly sick for only a short period before death. Symptoms are typical of a septicemic disease and include a general weakness, listlessness, lack of appetite and sometimes a yellowish or greenish diarrhea. Occasionally, the snood of toms may be turgid, swollen and purple. Some birds may be found lame with swollen leg joints due to localization of the infection. In breeding flocks, this disease occasionally is associated with decreased fertility and hatchability. Daily morbidity and mortality usually are low; however, in untreated flocks mortality may persist for some time and become excessive.

The most characteristic lesions are small or diffuse hemorrhages located in almost any tissue or organ. Such hemorrhages are commonly observed in the muscles, heart, liver, spleen, fat and other tissues of the body cavities. Hemorrhagic conditions of skin may result in purple blotches. The liver and spleen are usually enlarged, congested and occasionally contain necrotic foci. Enteritis or inflammation of the intestinal tract is commonly observed, as in most septicemic diseases.

Symptoms and lesions may resemble other diseases so closely that a reliable diagnosis can be made only through isolation and identification of the causative organism. Good management practices that aid in preventing erysipelas include avoiding the use of ranges previously occupied by swine, sheep or turkeys where erysipelas is known to have existed. Debeaking, removal of the snoods of toms, measures that prevent injury from fighting, avoiding overcrowding and providing well drained ranges will aid in preventing this disease problem. Bacterins are available and are useful on premises where history indicates that outbreaks may be a problem. The amount and duration of protection is relative to the amount of exposure and may not be sufficient for the entire laying period. Administer bacterins in accordance with the manufacturer's directions.

Move sick birds to a hospital pen for individual treatment and to prevent cannibalism. Moving unaffected birds to a clean range may aid in preventing the spread of the disease but may also contaminate an additional range. Various antibiotics have shown efficacy in treating erysipelas; however, penicillin is best. Penicillin injections in the leg or breast muscles of visibly sick birds is effective in decreasing mortality. One injection is usually sufficient, but more may be given if necessary. Water and feed medication may be of value under certain conditions.

Chapter

2

Soil Bacteria and Actinomycetes

What they are and a Few Interesting Facts

Bacteria are minuscule, one-celled organisms that can only be seen with a powerful light (1000X) or electron microscope (we're talking TINY). They can be so numerous that a pinch of soil can contain millions of organisms. Bacteria are tough—they occur everywhere on earth and have even been found over a mile down into the core of the earth. Bacteria can be classified into five functional groups. Autotrophic (literally, self-feeding) bacteria are photosynthetic. They are the primary producers. Decomposers consume soil organic matter, plant litter, and simple carbon compounds, releasing the nutrients in these substances for use by living plants. Mutualists, such as nitrogen-fixing bacteria, form associations with plants and help them absorb nutrients. Pathogens are the bad guys— they cause disease in plants. The last group, the chemolithotrophs (literally, chemical and rock-eating) obtain energy from minerals rather than from carbon compounds.

Bacteria are common throughout the soil, but tend to be most abundant in or adjacent to plant roots, an important food source.

Actinomycetes are a broad group of bacteria that form thread-like filaments in the soil. They are responsible for the distinctive scent of freshly exposed, moist soil.

Why they are Important

Bacteria are important in the carbon cycle. They contribute carbon to the system by fixation (photosynthesis) and decomposition. Bacteria are important decomposers in grassland environments. Actinomycetes are particularly effective at breaking down tough substances like cellulose (which makes up the cell walls of plants) and chitin (which makes up the cell walls of fungi) even under harsh conditions, such as high soil pH. Some management activities, particularly those that change nutrient levels in the soil, can shift the dominance of decomposers from bacterial to fungal. When one group becomes dominant where it shouldn't be, there is also a shift in the rest of the system. The shift from bacterial to fungal dominance, for instance, can enhance the conditions favoring weed invasions on rangelands.

Bacteria are particularly important in nitrogen cycling. Free-living bacteria fix atmospheric nitrogen, adding it to the soil nitrogen pool. Other nitrogen-fixing bacteria form associations with the roots of leguminous plants such as lupine, clover, alfalfa, and milkvetches. Actinomycetes form associations with some non-leguminous plants (important species are bitterbrush, mountain mahogany, cliffrose, and ceanothus) and fix nitrogen, which is then available to both the host and other plants in the near vicinity. Some soil nitrogen is unusable by plants until bacteria convert it to forms that can be easily assimilated. Some bacteria exude a sticky substance that helps bind soil particles into small aggregates. So despite their small size, they help improve water infiltration, water-holding capacity, soil stability, and aeration. Wait! Aren't there also "bad" bacteria? Yes, there are, but some soil bacteria suppress root-disease in plants by competing with pathenogenic organisms. The key is in maintaining a healthy system so that the good guys can do their work.

Bacteria are becoming increasingly important in bioremediation, meaning that we (people) can use bacteria to help us clean up our messes. Bacteria are capable of filtering and degrading a large variety of human-made pollutants in the soil and groundwater so that they are no longer toxic. The list of materials they can detoxify includes herbicides, heavy metals, and petroleum products.

Some Soil Bacteria Protect Soybeans from Root Diseases

Brian McSpadden Gardener, a plant pathologist with the Ohio Agricultural Research and Development Centre, is studying biological control of plant pathogens using beneficial bacteria that colonize plant roots. The goal of the research is to identify the distribution of soil bacteria across Ohio to determine which soybean fields would be ideal candidates for biological control applications.

"We know that some bacteria can promote plant growth but what we don't know is how much of those beneficial bacteria are in individual fields or how widely distributed they are across any defined geographical area," said McSpadden Gardener. "The ultimate goal is to find out if there is a way to assay a field and say okay, there are very low populations of beneficial bacteria here, so this would be an ideal spot to apply biological control."

In his work, in collaboration with Ohio State University Extension soybean agronomist Jim Beuerlein, McSpadden Gardener has identified two bacterial genotypes specific to Ohio soils that produce a natural fungicide known as DAPG (2,4-diacetylphloroglucinol). DAPG-producing bacteria that colonize soybean roots protect the plant from a wide range of root rot diseases, including Phytophthora, Pythium, Rhizoctonia and Fusarium.

"What we are doing is isolating different representatives of these DAPG-producing bacteria and treating seeds with them to see how well they colonize soybean roots and how well they protect roots against diseases," said McSpadden Gardener. "Our goal is to improve stand and increase plant yields by applying these DAPG-producing bacteria to seed as natural fungicides."

The DAPG compound works much like a chemical seed treatment fungicide in that it diffuses away from the plant roots, protecting the plant from fungal pathogens. But, unlike chemicals that lose their effectiveness once the plant grows out of its juvenile stage, DAPG-producing bacteria provide the plant protection from germination up to harvest.

"With synthetic chemicals, once that plant grows out of its sphere of influence, the fungi can sniff out the roots and attack," said McSpadden Gardener. "The idea behind biological control is that if you put the protective bacteria on the seed, as the plant germinates, the roots get colonized by the beneficial bacteria and they keep reproducing as the roots grow so more of the roots are protected more of the time. There's no other way to protect plants like that with a seed treatment."

During the two-year study, funded by an OARDC Seed Grant and the U.S. Department of Agriculture National Research Initiative, the researchers found that at 10 soybean sites throughout Ohio only about half of the plants harbored DAPG-producing bacteria. McSpadden Gardener said soil type, bacterium genotype, soybean variety, weather and production management are just some of the factors that determine the level of beneficial bacteria that are present to colonize plant roots.

"That's why it's so important to identify genotypes that are indigenous to Ohio," said McSpadden Gardener, who is also studying bacillus species that may prove effective in Ohio against soybean diseases. "A non-indigenous bacterium is not familiar with the Ohio soils or the environment, so it may not be as effective."

Seedling diseases cause a loss of about 3/4 of a million tons of seed annually, accounting for over $150 million in U.S. losses. Beneficial bacteria may be used as natural, effective and economical disease controls because of the unique relationship plants have with soil microbes. Between 5 and 10 percent of the carbon a plant fixes is released into the soil providing food for bacteria that colonize plant roots. Managing those plant-microbe associations can improve plant health.

Soil Bacteria's Role in Healthy Soil

Bacteria are the unsung heroes of gardening. Without bacteria, gardeners would not have the rich, loamy soil in which plants grow without bounds. Bacteria are single-celled organisms that chemically digest organic matter in soils into smaller nutrient components in forms available to plants. There are hundreds of thousands of different kinds of bacteria, and many types of bacteria can digest hundreds of different forms of organic matter into humus. The bacteria are able to do this because they can produce many different types of enzymes to digest different compounds.

Bacteria in Soil vs. Bacteria in Compost

Each type of bacteria works best under certain conditions. In a cool compost pile, or in garden soil, bacteria that thrive in cool conditions will form the bulk of the bacterial population. In hot compost, other bacteria that can survive the high temperatures dominate. When considering a bacterial soil conditioner, keep in mind the way in which you plan to use the conditioner. Understand the bacteria you are adding, if you are adding bacteria as soil conditioners, so that you add the right kind of bacteria. Once conditions change drastically, bacteria that cannot function in those conditions will perish. (Bacteria are not very mobile, unless they have water to float along in or wind to carry them.) The upside of bacterial immobility is that they will stay and multiply in place as long as conditions are favorable.

Bacteria for Soil Structure Improvement

Using a bacterial soil conditioning program not only helps with nutrient availability, it also helps to improve soil structure. Soils with poor structure benefit as bacteria breaks down soil compounds and the soil re-aggregates. Spaces for air and water will open up, and the structure of the soil will become more uniform.

Add Beneficial Bacteria for Oxygenation

Well structured soils provide plants with necessary oxygen in the root zone. Plants use carbon dioxide for photosynthesis, but they use oxygen for respiration, which is the process whereby plants break down stored sugars and starches to use as energy for growth. They get their oxygen by absorbing it in the root zone. Soils with good structure have plenty of spaces for oxygen. Soils without structure or organic matter generally do not have enough oxygen. When the plant cannot successfully undergo respiration, it cannot grow well.

Not all Bacteria is Good Bacteria

There are beneficial bacteria and there are bacteria that spread diseases. For bacteria to aid in soil conditioning, they need to be beneficial. Not just any bacteria will do. You can't really open your refrigerator, grab something old and throw it in the garden! You need the type of bacteria that will function well in your soil, in your climate, to decompose organic matter. Good bacteria does something other than break down organic matter. Good bacteria in soil is instrumental in keeping harmful fungi and viruses at bay. Be careful about your source for garden bacteria. Many will sing the praises of compost tea for adding bacteria to your soil, but these teas can also add pathogens that will harm your plants. If adding bacteria to your soil, it is best to go with a reputable source that can provide beneficial bacteria that will produce results.

Soil Components and Microorganisms

The oil is composed of five major components, such as inorganic matter, organic matter, soil air, soil water and soil organisms. The inorganic matter is derived from the weathering of parents rocks; air and water are derived from the atmosphere and are present in the soil; the organic matter is derived from the decomposition of organic matter; and the soil organisms constitute the most vital component of soil body. The soils, in fact are excellent culture media for the growth of various organisms. It is convenient to divide soil organisms into two major categories. First category refers to all those organisms which Can be studies satisfactorily only with a microscope; these are the microorganisms which include bacteria, cyanobacteria, microfungi, microalgae and protozoans. The second category refers to the remaining organism represented mostly by animals which can be studied by naked eye, and some are even quite large.

The greatest amount of microorganisms (1,000,000 per cu cm) is found generally in the top layer of the soil up to a depth of 5 to 15 cm. In deeper layers (1.5 to 5m) individual microbes are found. However, they have been discovered at a depth of 17.5 m in coal, oil, and artesian water. Due to nutrients and mixture in the soil, the number of microorganisms in 2 gm of soil reaches a colossal number from 200 million (bacteria) in clay to 5,000 million in black soil.

Bacteria That Infect Plants

Most people think of bacterial diseases affecting animals and humans but plants can also succumb to bacterial infections. Bacteria can enter through cuts our other areas of damage that occur in the leaves or stems due to wind bending the plant, objects hitting the plant, aphids sucking the plant sap or animals grazing on it. The symptoms then depend on which internal tissues and structures the infection takes a hold in.

Although thousands of species of bacteria can cause disease in animals and humans, a much smaller number, probably about one hundred, are able to infect plants and damage them. Bacterial plant diseases are generally due to bacilli, bacteria that are shaped by rods, some of which are gram negative bacteria and some of which are gram positive. The majority of plant diseases due to bacterial infection occur in parts of the world that have a tropical weather pattern.

How Bacteria Damage Plants

Once a species of pathogenic bacteria gains entry to a plant it causes one of four main problems that leads to a less than healthy plant. Some bacteria produce enzymes that break down the cell walls of plants anywhere in the plant. They do this to break open to cell to gain access to the nutrients inside but the plant cells affected die quickly, causing parts of the plant to start rotting. This is why plant diseases that develop due to cell wall degrading enzymes are generally known as a type of 'rot'.

Some bacteria produce toxins that are damaging to plant tissues generally, usually causing early death of the plant. Others produce large amounts of polysaccharide sugars that have long chains and are very sticky. As these travel in the water carrying vessels, the xylem, they block the narrow channels, preventing water getting from the plant roots up to the shoots and leaves, again causing rapid death of the plant. Finally, some bacteria produce proteins that mimic plant hormones. These lead to overgrowth of plant tissue and tumours form. These grow rapidly, taking up valuable nutrients and energy resources that the plant would otherwise use for its own tissue growth and it becomes weakened and susceptible to attack by other pathogens such as fungi.

Types of Bacteria That Infect Plants

One of the main types of bacteria that cause plant disease are the Proteobacteria. This large group of bacterial species include bacteria that cause disease in humans, such as Vibrio cholerae, which causes cholera, and Helicobacter pylori, which is associated with stomach ulcers. Some species of Proteobacteria live in soil where they fix nitrogen from the air, forming useful compounds that plants use in their growth. One species, Xanthomonas campestris, which affects plants from the Brassica and Arabidopsis groups causing black rot. The disease symptoms progress from discoloured leaves to extensive wilting of all of the stems and leaves, with all the leaves eventually turning yellow. Parts of the plant start to die and rot and the plant dies completely shortly afterwards.

Xanthomonas campestris causes its effects on the plant by producing large amounts of polysaccharides that block the xylem vessels that carry water through the plant. So much is produced that the bacteria is actually grown in culture to obtain these polysaccharides, better known as xanthan gum, a thickener used in the food industry.

Some species of Mycoplasma-like species of bacteria also cause serious plant disease. These are bacteria that do not have cell walls and they are introduced into the phloem tissue of plants by insects such as aphids that suck plant sap. Phloem vessels in plants are part of the transport system for sugars, which are generally moved from the leaves, where they are made by photosynthesis, to other parts of the plant, including the roots. Aphids use needle-like mouth parts to puncture plant stems and tap directly into this rich supply of liquid sugar. The Mycoplasma bacteria grow inside the aphids and other sap-sucking insects and are then introduced into their plant host by a very effective insect hypodermic. Plants that produce large amounts of sugar are the worst affected – for example coconut plants and sugar cane. The overall effect of the infection is to weaken the plant, opening it up to other diseases. The leaves usually lose their colour and another common symptom is green flowers, or an absence of flowers.

Plant Bacteria and Humans

Although some of the bacteria that infect plants are from the same large groups as bacterial pathogens in humans, there is no recorded instance of bacteria being passed from plant to person and causing an infection. Plant bacterial diseases are highly specific.

About Bacteria and Antibiotics

What are Bacteria?

Bacteria (singular: bacterium) are single-celled microscopic organisms that often live together in colonies. There are numerous types of bacteria. Most are harmless and some are even helpful, by aiding digestion or breaking down rotting material. A few bacteria can cause diseases in humans, animals and plants; these are referred to as pathogenic bacteria or pathogens.

Where do Bacteria Live?

Excepting normally sterile sites such as the blood stream and spinal fluid, bacteria are commonly found almost anywhere. Their natural habitats are soil, plants, water, and on the skin and in the digestive tracts of humans and animals. They may survive for varying amounts of time on other surfaces that have had contact with these environments.

What is an Antibiotic?

An antibiotic is a chemical compound that can kill or inhibit the growth of a bacteria without causing the host (such as a human or an animal) significant damage. Antibiotics can be naturally produced (like penicillin) by a mold or bacterium, or synthetically made (like the fluoroquinolones).

How do Antibiotics Work?

The activity of antibiotics may be classified as "bacteriostatic" or "bacteriocidal." Bacteriostatic drugs stop the cells from growing and multiplying, but do not actually kill them. This inhibition of growth allows time for the infected host to mount an immune response and eliminate the non-growing infectious agent. Bacteriocidal drugs actually kill the bacteria. These drugs are the preferred choice when a person's immune system is weakened and incapable of destroying the organism itself. In either case, the antibiotics act by disrupting bacterial cell activity, for example, by attacking the outer bacterial cell wall or its inner membrane, or the chemical pathways necessary for the bacteria to live and reproduce. Antibiotics have no activity against viruses.

How are Antibiotics used?

Antibiotics are used in human and veterinary medicine to treat illnesses caused by bacteria and, in some cases, to prevent bacterial infections (prior to surgery, for example). Antibiotics are also used to increase growth in certain food animals, and are used as pesticides to control bacterial infections in food crops.

How Many Antibiotics are There?

There are hundreds of antibiotics, most of which belong to eight major family groups. The drugs are classified according to their basic chemical structure. Most of the members within a family arise from additions or substitutions of attachments to the drug's core structure.

Antibiotic Resistance

What is Antibiotic Resistance?

Antibiotic resistance occurs when an antibiotic has lost its ability to effectively control or kill bacterial growth; in other words, the bacteria are "resistant" and continue to multiply in the presence of therapeutic levels of an antibiotic.

Why do Bacteria become Resistant to Antibiotics?

Antibiotic resistance is a natural phenomenon. When an antibiotic is used, bacteria that can resist that antibiotic have a greater chance of survival than those that are "susceptible." Susceptible bacteria are killed or inhibited by an antibiotic, resulting in a selective pressure for the survival of resistant strains of bacteria. Some resistance occurs without human action, as bacteria can produce and use antibiotics against other bacteria, leading to a low-level of natural selection for resistance to antibiotics. However, the current higher-levels of antibiotic-resistant bacteria are attributed to the overuse and abuse of antibiotics. In some countries and over the Internet, antibiotics can be purchased without a doctor's prescription. Patients sometimes take antibiotics unnecessarily, to treat viral illnesses like the common cold.

How do Bacteria become Resistant?

Some bacteria are naturally resistant to certain types of antibiotics. However, bacteria may also become resistant in two ways: 1) by a genetic mutation or 2) by acquiring resistance from another bacterium. Mutations, rare spontaneous changes of the bacteria's genetic material, are thought to occur in about one in one million to one in ten million cells. Different genetic mutations yield different types of resistance. Some mutations enable the bacteria to produce potent chemicals (enzymes) that inactivate antibiotics, while other mutations eliminate the cell target that the antibiotic attacks. Still others close up the entry ports that allow antibiotics into the cell, and others manufacture pumping mechanisms that export the antibiotic back outside so it never reaches its target.

Bacteria can acquire antibiotic resistance genes from other bacteria in several ways. By undergoing a simple mating process called "conjugation," bacteria can transfer genetic material, including genes encoding resistance to antibiotics (found on plasmids and transposons,) from one bacterium to another. Viruses are another mechanism for passing resistance traits between bacteria. The resistance traits from one bacterium are packaged into the head portion of the virus. The virus then injects the resistance traits into any new bacteria it attacks. Bacteria also have the ability to acquire naked, "free" DNA from their environment. Any bacteria that acquire resistance genes, whether by spontaneous mutation or genetic exchange with other bacteria, have the ability to resist one or more antibiotics. Because bacteria can collect multiple resistance traits over time, they can become resistant to many different families of antibiotics.

How does Antibiotic Resistance Spread?

Genetically, antibiotic resistance spreads through bacteria populations both "vertically," when new generations inherit antibiotic resistance genes, and "horizontally," when bacteria share or exchange sections of genetic material with other bacteria. Horizontal gene transfer can even occur between different bacterial species. Environmentally, antibiotic resistance spreads as bacteria themselves move from place to place; bacteria can travel via airplane, water and wind. People can pass the resistant bacteria to others; for example, by coughing or contact with unwashed hands.

Can Bacteria Lose their Antibiotic Tesistance?

Yes, antibiotic resistance traits can be lost, but this reverse process occurs more slowly. If the selective pressure that is applied by the presence of an antibiotic is removed, the bacterial population can potentially revert to a population of bacteria that responds to antibiotics.

Antibiotic Resistance: A Societal Problem

How can Antibiotic Resistance Harm Humans?

If large numbers of bacteria are resistant to antibiotics, it will be more difficult and more expensive to treat human bacterial infections. When antibiotics fail to work, consequences include extra visits to the doctor, hospitalization or extended hospital stays, a need for more expensive antibiotics to replace the older ineffective ones, lost workdays and, sometimes, death.

Why is Antibiotic Resistance a Public Health Problem?

Antibiotics are called "societal drugs," since antibiotic resistance can pass from bacterium to bacterium, and resistant bacterial infections can pass from person to person. Thus, antibiotic use and antibiotic resistance can eventually affect an entire community.

Why is Antibiotic Resistance an Ecological Problem?

When antibiotics are used in humans or animals, approximately 80-90% of the ingested antibiotics are not broken down, but pass through the body intact and enter the environment as waste. Thus, they retain their ability to affect bacteria and promote antibiotic resistance even after they enter the soil or water as a waste product.

Do People Become resistant to Antibiotics?

No, this is a common misconception. People may exhibit allergic reactions to antibiotics, but they are not resistant to them. It is the bacteria themselves, not the infected host, which become resistant.

How Serious is the Problem of Antibiotic Resistance?

Antibiotic resistance is found all over the world and has become a very serious problem in the treatment of disease. The 1995 US Office of Technology Assessment report attributed a cost of $1.3 billion (1992 dollars) per year for antibiotic-resistant infections due to six species of bacteria in US hospitals. While the real magnitude of the problem is unknown, the monetary cost of treating antibiotic resistant infections worldwide is estimated to be many billions of dollars per year. Some experts predict that, as resistance to antibiotics is increasing at a faster pace than it can be controlled, the future will resemble the preantibiotic era. Others are more optimistic that research and careful drug management can reverse the trend if global efforts are focused on recognizing and controlling it.

Bacteria as Plants

During the last fifteen years the subject of bacteriology has developed with a marvellous rapidity. At the beginning of the ninth decade of the century bacteria were scarcely heard of outside of scientific circles, and very little was known about them even among scientists. Today they are almost household words, and everyone who reads is beginning to recognise that they have important relations to his everyday life. The organisms called bacteria comprise simply a small class of low plants, but this small group has proved to be of such vast importance in its relation to the world in general that its study has little by little crystallized into a science by itself. It is a somewhat anomalous fact that a special branch of science, interesting such a large number of people, should be developed around a small group of low plants.

The importance of bacteriology is not due to any importance bacteria have as plants or as members of the vegetable kingdom, but solely to their powers of producing profound changes in Nature. There is no one family of plants that begins to compare with them in importance. It is the object of this work to point out briefly how much both of good and ill we owe to the life and growth of these microscopic organisms.

As we have learned more and more of them during the last fifty years, it has become more and more evident that this one little class of microscopic plants fills a place in Nature's processes which in some respects balances that filled by the whole of the green plants. Minute as they are, their importance can hardly be overrated, for upon their activities is founded the continued life of the animal and vegetable kingdom. For good and for ill they are agents of neverceasing and almost unlimited powers.

Historical

The study of bacteria practically began with the use of the microscope. It was toward the close of the seventeenth century that the Dutch microscopist, Leeuwenhoek, working with his simple lenses, first saw the organisms which we now know under this name, with sufficient clearness to describe them. Beyond mentioning their existence, however, his observations told little or nothing. Nor can much more be said of the studies which followed during the next one hundred and fifty years. During this long period many a microscope was turned to the observation of these minute organisms, but the majority of observers were contented with simply seeing them, marvelling at their minuteness, and uttering many exclamations of astonishment at the wonders of Nature. A few men of more strictly scientific natures paid some attention to these little organisms. Among them we should perhaps mention Von Gleichen, Muller, Spallanzani, and Needham. Each of these, as well as others, made some contributions to our knowledge of microscopical life, and among other organisms studied those which we now call bacteria. Speculations were even made at these early dates of the possible causal connection of these organisms with diseases, and for a little the medical profession was interested in the suggestion. It was impossible then, however, to obtain any evidence for the truth of this speculation, and it was abandoned as unfounded, and even forgot-ten completely, until revived again about the middle of the 19th century. During this century of wonder a sufficiency of exactness was, how-ever, introduced into the study of microscopic organisms to call for the use of names, and we find Müller using the naines of Monas, Proteus, Vibrio, Bacillus, and Spirillum, names which still continue in use, although commonly with a different significance from that given them by Müller. Muller did indeed make a study sufficient to recognise the several distinct types, and attempted to classify these bodies. They were not regarded as of much importance, but simply as the most minute organisms known.

Nothing of importance came from this work, however, partly because of the inadequacy of the microscopes of the day, and partly because of a failure to understand the real problems at issue. When we remember the minuteness of the bacteria, the impossibility of studying any one of them for more than a few moments at a time—only so long, in fact, as it can be followed under a microscope; when we remember, too, the imperfection of the compound microscopes which made high powers practical impossibilities; and, above all, when we appreciate the looseness of the ideas which pervaded all scientists as to the necessity of accurate observation in distinction from inference, it is not strange that the last century gave us no knowledge of bacteria beyond the mere fact of the existence of some extremely minute organisms in different decaying materials. Nor did the 19th century add much to this until toward its middle. It is true that the microscope was vastly improved early in the century, and since this improvement served as a decided stimulus to the study of microscopic life, among other organisms studied, bacteria received.

Some attention. Ehrenberg, Dujardin, Fuchs, Perty, and others left the impress of their work upon bacteriology even before the middle of the century. It is true that Schwann shrewdly drew conclusions as to the relation of microscopic organisms to various processes of fermentation and decay—conclusions which, although not accepted at the time, have subsequently proved to be correct. It is true that Fuchs made a careful study of the infection of " blue milk," reaching the correct conclusion that the infection was caused by a microscopic organism which he discovered and carefully studied.

It is true that Henle made a general theory as to the relation of such organisms to diseases, and pointed out the logically necessary steps in a demonstration of the causal connection between any organism and a disease. It is true also that a general theory of the production of ail kinds of fermentation by living organisms had been advanced. But all these suggestions made little impression. On the one hand, bacteria were not recognised as a class of organisms by themselves-were not, indeed, distinguished from yeasts or other minute animalculæ.

Their variety was not mistrusted and their significance not conceived. As microscopic organisms, there were no reasons for considering them of any more importance than any other small animals or plants, and their extreme minuteness and simplicity made them of little interest to the microscopist. On the other hand, their causal connection with fermentative and putrefactive processes was entirely obscured by the overshadowing weight of the chemist Lie-big, who believed that fermentations and putrefactions were simply chemical processes. Liebig insisted that all albuminoid bodies were in a state of chemically unstable equilibrium, and if left to themselves would fall to pieces without any need of the action of microscopic organisms.

The force of Liebig's authority and the brilliancy of his expositions led to the wide acceptance of his views and the temporary obscurity of the relation of microscopic organisms to fermentative and putrefactive processes. The objections to Liebig's views were hardly noticed, and the force of the experiments of Schwann was silently ignored. Until the sixth decade of the century, therefore, these organisms, which have since become the basis of a new branch of science, had hardly emerged from obscurity. A few microscopists recognised their existence, just as they did any other group of small animals or plants, but even yet they failed to look upon them as forming a distinct group. A growing number of observations was accumulating, pointing toward a probable causal connection between fermentative and putrefactive processes and the growth of microscopic organisms; but these observations were known only to a few, and were ignored by the majority of scientists.

It was Louis Pasteur who brought bacteria to the front, and it was by his labours that these organisms were rescued from the obscurity of scientific publications and made objects of general and crowning interest. It was Pasteur who first successfully combated the chemical theory of fermentation by showing that albuminous matter had no inherent tendency to decomposition. It was Pasteur who first clearly demonstrated that these little bodies, like all larger animals and plants, come into existence only by ordinary methods of reproduction, and not by any spontaneous generation, as had been earlier claimed. It was Pasteur who first proved that such a common phenomenon as the souring of milk was produced by microscopic organisms growing in the milk. It was Pasteur who first succeeded in demonstrating that certain species of microscopic organisms are the cause of certain diseases, and in suggesting successful methods of avoiding them. All these discoveries were made in rapid succession. Within ten years of the time that his name began to be heard in this connection by scientists, the subject had advanced so rapidly that it had become evident that here was a new subject of importance to the scientific world, if not to the public at large. The other important discoveries which Pasteur made it is not our purpose to mention here. His claim to be considered the founder of bacteriology will be recognised from what has already been mentioned. It was not that he first discovered the organisms, or first studied them; it was not that he first suggested their causal connection with fermentation and disease, but it was because he for the first time placed the subject upon a firm foundation by proving with rigid experiment some of the suggestions made by others, and in this way turned the attention of science to the study of microorganisms.

After the importance of the subject had been demonstrated by Pasteur, others turned their attention in the same direction, either for the purpose of verification or refutation of Pasteur's views. The advance was not very rapid, however, since bacteriological experimentation proved to be a subject of extraordinary difficulty. Bacteria were not even yet recognised as a group of organisms distinct enough to be grouped by themselves, but were even by Pasteur at first confounded with yeasts. As a distinct group of organisms they were first distinguished by Hoffman in 1869, since which date the term bacteria, as applying to this special group of organisms, has been coming more and more into use. So difficult were the investigations, that for years there were hardly any investigators besides Pasteur who could successfully handle the subject and reach conclusions which could stand the test of time. For the next thirty years, although investigators and investigations continued to increase, we can find little besides dispute and confusion along this line. The difficulty of obtaining for experiment any one kind of bacteria by itself, unmixed with others (pure cultures), rendered advance almost impossible. So conflicting were the results that the whole subject soon came into almost hopeless confusion, and very few steps were taken upon any sure basis. So difficult were the methods, so contradictory and confusing the results, because of impure cultures, that a student of today who wishes to look up the previous discoveries in almost any line of bacteriology need hardly go back of 1880, since he can almost rest assured that anything done earlier than that was more likely to be erroneous than correct.

The last fifteen years have, however, seen a wonderful change. The difficulties had been mostly those of methods of work, and with the ninth decade of the century these methods were simplified by Robert Koch. This simplification of method for the first time placed this line of investigation within the reach of scientists who did not have the genius of Pasteur. It was now possible to get pure cultures easily, and to obtain with such pure cultures results which were uniform and simple. It was now possible to take steps which had the stamp of accuracy upon them, and which further experiment did not disprove. From the time when these methods were thus made manageable the study of bacteria increased with a rapidity which has been fairly startling, and the information which has accumulated is almost formidable. The very rapidity with which the investigations have progressed has brought considerable confusion, from the fact that the new discoveries have not had time to be properly assimilated into knowledge. Today many facts are known whose significance is still uncertain, and a clear logical discussion of the facts of modern bacteriology is not possible. But sufficient knowledge has been accumulated and digested to show us at least the direction along which bacteriological advance is tending, and it is to the pointing out of these directions that the following pages will be devoted.

Industrial Enzymes Derived from Plants, Vegetables, Fungus and Bacteria

Enzymes

Alpha Acetolactate De-carboxilase: Alpha-Acetolactate decarboxylase is used to avoid formation of the unpleasant tasting alpha-diacetyl from alpha-acetolactate during fermentation. In the traditional brewing process, the alpha-diacetyl formed from alpha-acetolactate is further reduced to acetoin over a 2-to 4-week maturation period alpha-Acetolactate decarboxylase causes direct decarboxylation of alpha-acetolactate to acetoin, thus avoiding the need for this maturation period. The enzyme can similarly be used in the fermentation of alcohol, where diacetyl is otherwise formed and requires a maturation step before distillation.

Amyloglucosidase

Amyloglucosidase (EC 3.2.1.3) is an important industrial enzyme that is used by the manufacturers of high fructose corn syrup (HFCS). More than half of all commercial baked goods and practically all soft drink bottlers use this syrup instead of sugar because HFCS is both sweeter and cheaper. Amyloglucosidase catalyzes the breakdown of malto-oligosaccharides to glucose.

Bacterial Alpha Amylase

Amylase is a digestive enzyme classified as a saccharidase (an enzyme that cleaves polysaccharides). It is mainly a constituent of pancreatic juice and saliva, needed for the breakdown of long-chain carbohydrates (such as starch) into smaller units. Amylase is a digestive enzyme made primarily by the pancreas and salivary glands. Enzymes are substances made and used by the body to trigger specific chemical reactions. The primary function of the enzyme amylase is to break down starches in food so that they can be used by the body.

Barley Beta-Amylase

Beta-amylase is a 1,4-alpha-D-glucan maltohydrolase (I.U.B. 3.2.1.2) and is present in ungerminated and germinated (malted) barley. Beta-amylase occurs as several isozymes which are immunologically identical. Beta-amylase in ungerminated barley is less soluble in water and migrates more rapidly at pH 8.2 than that of malted barley, but that the two are antigenically identical.

Beta-Glucanase

The glucanase will attack on the cellulose fibre to liberate smaller fragments of cellulose which is further attacked by exo-cellulase to liberate glucose.

Catalase

The ability of organisms to use molecular oxygen was a major evolutionary breakthrough that enabled the production of significantly more energy from the breakdown of foods, amongst many other advantages. However, these advantages came at a cost: toxic by-products known as reactive oxygen species (ROS) are produced, which if left unchecked would seriously affect an organism's viability. These ROS include hydrogen peroxide, superoxide anion radicals, singlet oxygen, hydroxyl radicals and nitric oxide.

Cellulase

Cellulase is an enzyme complex which breaks down cellulose to beta-glucose. It is produced mainly by symbiotic bacteria in the ruminating chambers of herbivores. Aside from ruminants, most animals (including humans) do not produce cellulase, and are therefore unable to use most of the energy contained in plant material

Cellulase refers to a family of enzymes which act in concert to hydrolyze cellulose. Cellulases are widely distributed throughout the biosphere and are most manifest in fungal and microbial organisms.

Coenzyme Q10

Coenzyme Q10 (CoQ 10) or ubiquinone is essentially a vitamin or vitamin-like substance. Disagreements on nomenclature notwithstanding, vitamins are defined as organic compounds essential in minute amounts for normal body function acting as coenzymes or precursors to coenzymes. They are present naturally in foods and sometimes are also synthesized in the body. CoQ10 likewise is found in small amounts in a wide variety of foods and is synthesized in all tissues.

Dextranase

Dextran is a chemically and physically complex polymer, breakdown of which is carried out by a variety of endo-and exodextranases. Enzymes in many groups can be classified as dextranases according to function: such enzymes include dextranhydrolases, glucodextranases, exoisomaltohydrolases, exoisomaltotriohydrases, and branched-dextran exo-1,2--glucosidases. Cycloisomalto-oligosaccharide glucanotransferase does not formally belong to the dextranases even though its side reaction produces hydrolyzed dextrans. A new classification system for glycosylhydrolases and glycosyltransferases, which is based on amino acid sequence similarities, divides the dextranases into five families.

Fungal Alpha Amylase

The enzyme a-amylase from Aspergillus oryzae used in bakeries to improve the bread quality has been identified as an inhalative allergen in baker's asthma. It is doubtful whether this enzyme can induce allergic sensitization in regular bread consumers.

The monitoring of the fungal and wheat alpha-amylase activity is of great importance to grain elevators, grain inspectors, flour mills, wheat growers, fungal alpha-amylase manufacturers, and bakeries. Produced by Aspergillus oryzae and Aspergillus niger mold, fungal alpha-amylase is added to flour by bakers to increase the quality of baked goods. Wheat alpha-amylase naturally occurs in wheat that is exposed to excessive moisture in the fields. Bakers and millers need to quickly and easily determine the exact level of wheat alpha-amylase in flour and fungal alpha-amylase in the additive.

Enzymes I

Fungal Diastase

The Fungal Diastase Or Fungal Alpha Amylase is a potent, dependable, powerful enzyme complex, obtained from Aspergillus oryzae. The Fungal Diastase is manufactured by controlled fermentation of Fungi Aspergillus oryzae, followed by a sophisticated down stream process of recovery.

Fungal Lactase

Lactase is an enzyme in the small intestine that digests lactose (the naturally occurring sugar in milk). Normally lactase breaks lactose down into a form that can be easily absorbed by the blood. Milk, ice cream, yogurt and other dairy products have varying levels of lactose, which affect how much lactase is required for proper digestion.

Fungal Lipase

Fungal Lipase is manufactured from genus aspergillus by fermentation. This is followed by sophisticated down stream processes, which includes filtration, concentration, micro filtration and finally spray drying to fine powder.

Glucoamylase

Glucoamylase (EC 3.2.1.3) is an important industrial enzyme that is used by the manufacturers of high fructose corn syrup (HFCS). More than half of all commercial baked goods and practically all soft drink bottlers use this syrup instead of sugar because HFCS is both sweeter and cheaper. Glucoamylase catalyzes the breakdown of malto-oligosaccharides to glucose.

Amylase is a digestive enzyme classified as a saccharidase (an enzyme that cleaves polysaccharides). It is mainly a constituent of pancreatic juice and saliva, needed for the breakdown of long-chain carbohydrates (such as starch) into smaller units.

Glucose Oxidase

Glucose oxidase (S-?D-glucose: oxygen 1-oxidoreductase, EC1.1.3.4) catalyses the oxidation of S-D-glucose to D-glucono-1,5-lactone and hydrogen peroxide, using molecular oxygen as the electron acceptor.

Glucose oxidase is widely used for the determination of glucose in body fluids and in removing residual glucose and oxygen from beverages and foodstuffs. Furthermore, Glucose oxidase-producing moulds such as Aspergillus and Penicillum Species are used for the biological production of gluconic acid.

Hemicellulase

Hemicellulase is an enzyme that breaks down hemicellulose. Hemicellulose catagorizes a variety of polysaccharides that are more complex than sugars and less complex than cellulose, that are found in plant walls.

It is a mixture of enzymes which can hydrolyze the indigestible components of plant fibres. Since humans lack the endogenous enzymes required to digest plant fibres, the supplementation of hemicellulase provides humans with an additional source of nutrition and reduces the bulking effect of fibrous foods. Scientific evidence suggests that carbohydrolytic enzymes, such as Hemicellulase, can be useful supplements for digestive support and general nutritional support.

Invertase

Invertase is a yeast-derived enzyme. Invertase splits sucrose into glucose and fructose (invert syrup) and can be applied for any inversion of sucrose especially liquefied cherry centres, creams, mints, truffles, marshmallow, invert syrup and other fondants. Invertase is used to improve shelf life of confections.

The official name for invertase is beta-fructofuranosidase (EC3.2.1.26), which implies that the reaction catalyzed by this enzyme is the hydrolysis of the terminal nonreducing beta-fructofuranoside residues in beta-fructofuranosides. Note that alpha-D-glucosidase, which splits off a terminal glucose unit, can also catalyze this reaction. Note that sucrose can be hydrolyzed relatively easily; the reaction proceeds in an acidic environment without the aid of invertase.

Laccase

Laccase is a copper-containing enzyme, 1,4-benzenediol oxidase (EC 1.10.3.2), found in higher plants and microorganisms. Laccases are multicopper oxidases of wide specificity that carry out one-electron oxidation of phenolic and related compounds, and reduce O2 to water. The enzymes are polymeric and generally contain one each of type 1, type 2, type 3 copper centres per subunit, where the type 2 and type 3 are close together forming a trinuclear copper cluster.

Malt Diastase

Malt diastase is a carbohydrolytic enzyme useful for digestive support and general nutrition support. Malt diastase is characterized by the ability to break down amylose and other polysaccharides. The enzyme works with amylase and glucoamylase to digest carbohydrate rich foods such as grains as well as malt, maltose, and sugars. Malt diastase is also known as maltase. It is produced by the cells lining the small intestine.

Nattokinase

Nattokinase is a potent fibrinolytic enzyme extracted and highly purified from a traditional Japanese food called Natto. Natto is a fermented cheese-like food that has been used in Japan for over 1000 years for its popular taste and as a folk remedy for heart and vascular diseases.

Natto is produced by a fermentation process by adding Bacillus natto, a beneficial bacteria, to boiled soybeans. The resulting Nattokinase enzyme, is produced when Bacillus natto acts on the soybeans. While other soy foods contain enzymes, it is only the natto preparation that contains the specific nattokinase enzyme.

Neutral Cellulase

The activity of any enzyme depends primarily on the pH and temperature of the reaction system. Basically, cellulases can be divided in to four basic groups according to the pH required for optimum enzymatic activity.

One group is called Acid Cellulases because the pH ranges from 4.5 to 5.0. There are Hybrid Cellulases whose optimum pH is 4.5 to 7.0. The third group called the Neutral Cellulases is active from a pH 6.0 to 8.0 but is optimal at pH 6.2. The fourth group is commonly referred as to Alkaline Cellulase enzymes because its optimum pH is 7.2 to 8.5.

Enzymes II

Ox Bile Extract

Bile is a secretion of liver, stored in the gall bladder until needed. Bile has a major role to play in fat digestion and absorption. The active constituents of bile are the bile salts, which are either glycine or taurine conjugates of polyhydroxy steroidal acid. Principal bile acids of mammals are the hydroxyl derivatives of 5-beta-cholan-24 oic-acid. Ox-bile is the most important commercial source of these acids and contains primarily Cholic acid with less amount of deoxycholic acid.

Pancreatin

Pancreatin is a digestive enzyme that is used to supplement loss of or low levels of digestive enzymes, and is often used in people with cystic fibrosis. It has been claimed to help with food allergies, celiac disease, autoimmune disease, cancer, and weight loss. Pancreatin is also known as pancreatic acid.

Papain

Papain is naturally occurring enzyme, obtained from latex of fruit of PAPAYA (Carica papaya Latex). The enzyme Papain is obtained by drying papaya latex by suitable method viz; sun drying, tray drying, spray drying etc. spray drying gives a fine powder, which dissolves easily in water to provide a solution of high proteolytic activity.

Pectinase

Pectinase is a general term for enzymes that break down pectin, a polysaccharide substrate that is found in the cell walls of plants. One of the must studied and widely used commercial pectinases is polygalacturonase. It is useful because pectin is the jelly-like matrix which helps cement plant cells together and in which other cell wall components, such as cellulose fibrils, are embedded. Therefore pectinase enzymes are commonly used in processes involving the degradation of plant materials, such as speeding up the extraction of fruit juice from fruit, including apples. Pectinases have also been used in wine production since the 1960s.

Pepsin

Pepsin is one of the principal protein degrading or proteolytic enzymes in the digestive system. During the process of digestion, Pepsin acts on the complex dietary protein and breaks up into peptides and amino acids which can be readily absorbed by the intestinal lining.

Peroxidase

Peroxidases are a group of enzymes that catalyze oxidation-reduction reactions. As such, they are classified as oxidoreductases. Toxic molecules such as superoxide and hydroxide radicals can be found in cells due to the presence of oxygen. These are byproducts of aerobic respiration. They are eliminated by a number of enzymes present inside the cell. Superoxide, for example, is destroyed by superoxide dismutase. The degradation, however, produces more hydrogen peroxide (H2O2), which is, in turn, destroyed by peroxidase. Peroxidases reduce H2O2 to water while oxidizing a variety of substrates. Thus, peroxidases are oxidoreductases which use H2O2 as electron acceptor for catalyzing different oxidative reactions.

Phytase

Large amount of research show that phytate or phytic acid in monogastric animal's feeds has a strong anti-nutrition effect the reason being: Monogastric animals lack phytate degrading enzyme in their digestive tract so they are unable to utilize the phosphorus of phytate

Protease

Proteases (Proteinases, Peptidases or Proteolytic enzymes) are enzymes that break peptide bonds between amino acids of proteins. The process is called proteolytic cleavage, a common mechanism of activation or inactivation of enzymes especially involved in blood coagulation or digestion. They use a molecule of water for this and are thus classified as hydrolases.

Pullulanase

Pullulanase is a specific kind of glucanase, an amylolytic exoenzyme, that degrades pullulan. It is produced as an extracellular, cell surface-anchored lipoprotein by Gram-negative bacteria of the genus Klebsiella. Type I pullulanases specifically attack a-1,6 linkages, while type II pullulanases are also able to hydrolyse a-1,4 linkages. It is also produced by some other bacteria and archaea. Pullanase is used as a detergent in biotechnology.

Serratiopeptidase

Serratiopeptidase is derived from bacteria belonging to the genus Serratia. The controlled fermentation of Serratia sp. secretes this enzyme in the highly selective medium. The recovery process involves various types of filtration, concentration and steps to make enzyme useful for pharmaceutical applications and finally dried to fine free flowing powder form.

Sucrase

Sucrase is a yeast-derived enzyme. Sucrase splits sucrose into glucose and fructose (invert syrup) and can be applied for any inversion of sucrose especially liquefied cherry centres, creams, mints, truffles, marshmallow, invert syrup and other fondants. Sucrase is used to improve shelf life of confections. It is available in single, double and triple strengths and is packaged in one, ten and 44 pound containers for ease of use, storage and cost efficiency.

Trypsin Chymotrypsin Mix

Trypsin Chymotrypsin Mix is a unique anti-inflammatory product manufactured by Advanced Enzyme Technologies. Trypsin Chymotrypsin Mix is a proteolytic enzyme obtained by controlled proteolysis of zymogens form in the frozen glands. Successive conc. and purification steps give Trypsin Chymotrypsin Mix best suitable for pharmaceutical applications.

Xylanase

Xylanase produced by submerged fermentation of Trichoderma reesei var. It has cellulase side activity. This is classified by Nomenclature Committee of the International Union of Biochemistry and Molecular Biology (NC-IUBMB) in consultation with the IUPAC-IUBMB joint Commission on Biochemical Nomenclature (JCBN).

Chapter

3

Canine Bacterial Diseases

Canine bacterial diseases are much different than viral infections, although both can be very serious threats to your dog. In fact, one of the oldest and most destructive diseases in recorded history not only for people but dogs is a bacterial disease called anthrax. In fully understanding bacterial infections in dogs, in helps to understand the differences between the types of germs that are the major causes of various diseases in your pet. The term germ is a reference to any microorganism. Not all microorganisms cause diseases, but those that do will either be a bacterium, a virus, or some type of fungi. They are often confused and grouped together, but they are very, very different.

Bactria is different for a virus in that they are much larger. In fact, the largest virus is just barely as big as the smallest bacterium. However, they are still very small and usually can not be seen by the human eye; but they are much more complex than a virus.

Most all forms of bacteria have an outer cell wall that is very rigid and come in three basic shapes; spheres, rods, or spirals. They also have a lining on the inside of the cell that is a plasma type membrane. This is important because it provides for the support of the bacterium, allowing it to control its environment. On the very inside of the structure of bacteria is what is referred to as ribosomes. All living cells have this same substance, but in bacteria it is smaller that what is found in a regular cell. This is an extremely important point to understand as most modern day antibiotics as well as some very old natural treatments, attack this substance directly. Because it is smaller, antibiotics can find it and destroy the bad ones while the good ones in normal cells are left alone. If the immune system in your dog is strong enough, most of these bacteria do not make it past the immune attack points. However, if it is weak or compromised, your dog's immune system is breached and as a result these attacking bacteria can cause several types of diseases.

Canine Bacterial Diseases

Lyme Disease

This infectious disease in dogs is caused by a bacteria called a spirochete and is transmitted by ticks. The most common tick to carry this disease is a deer tick and is found primarily in the Northeast, the Upper Midwest, and in all Western states. For your dog to become infected, this tick must be attached to your pet for at least 48 hours. If it is found and removed before that time your dog will not catch this potentially deadly disease.

What makes this disease especially dangerous is that if the tick does drop off by accident, they seek another host as soon as possible, including humans. The symptoms of this disease in your dog will usually not surface for two to three months, but when they do, they are severe. First will be a very high temperature, usually about 103 degrees, followed by lameness and joint swelling. If severe, it could spread and attack your dog's kidneys, causing kidney failure. There is a vaccination for this disease, but is highly controversial as to how effective it actually is.

Leptospirosis

This bacterial disease is found worldwide and is especially vicious as there are several different stains of these bacteria. Although there is a vaccine for this disease, it is just like the flu in that it is only good for that one particular strain. It is caught usually by infected urine, placental transfers, or bites. Stagnate or very slow moving water can also spread this bacteria. The symptoms include a very high fever of 103 degrees plus, shivering, and muscle tenderness. As it progresses, your dog will start to vomit excessively and dehydrate very rapidly. Severely infected dogs may than contract hypothermia, which can be fatal.

Brucellosis

This disease is caused by bacteria that are spread from dead puppy fetuses. It is estimated that ten percent of all dogs carry this bacteria. It is the most common bacteria that will affect bitch dogs and their fetuses. It is especially dangerous as it can also be transmitted to humans. Transmission is caused by the sexual mating of infected dogs, and this bacterium can live in the vaginal and uterine tissues for years, and in some cases, for life. It can than be passed onto the offspring causing the disease. If the infection is severe, it causes the litter to be aborted or the puppies to die shortly after birth.

If the embryos die in the early stages, they may be absorbed back into the female without ever showing signs of being pregnant. There is no form of reliable treatment at this time for this disease. There are a few other diseases that are caused by what is referred to as rickettsia, which is technically somewhere between a bacterium and a virus on the evolutionary scale. They include Ehrlichiosis and Rocky Mounted Spotted Fever.

Preventions

So how exactly does your dog fight off these bacterial infections? Your dog will produce antibodies that will fight these infections, but it usually takes about a week for the process to consummate. While that is happening, antibiotics are used if the disease is caught early enough. However, there is a growing concern that antibiotics are used too frequently in dogs and can actually kill the friendly bacteria in your dog. These friendly bacteria help to produce natural vitamins that help host organisms to fight off these infections. In fighting bacteria in dogs, there are several natural remedies that can assist your pet. Phytonutrients are natural occurring compounds that are found in plants. They have a tremendous impact on the immune systems in your dog as they are have natural antioxidant and antibacterial affects without harming the friendly bacteria. They also help the cells protect against oxidation damage and assist in the normal process protecting against invading ribosomes.

Bromelain, which is a digestive enzyme that comes from pineapples, has proven very effective in animal studies in protecting against bacteria's ability to adhere to the mucosal lining in the digestive wall. This is bacteria's biggest strength, and if this is broken down it has very little chance of forming. Ginger, an herb found in ginger root has also been found to be very affective in inhibiting both gram-positive and gram-negative bacteria.

The Nitrogen Cycle

One of the most common and freely available elements on Earth is nitrogen. The air that we breathe comprises of about 78% nitrogen and it appears in different forms throughout life's various biological cycles. Being an enclosed system, an aquarium requires careful monitoring to ensure that harmful forms of nitrogen do not impact negatively on it's occupants. An understanding of how the Nitrogen cycle works is an essential component to successful fish keeping. The chemical reactions which take place in nature are complex and an aquarium is only a reconstruction of a part of the natural environment. The scale may be smaller, but essentially the fundamentals remain the same.

Hopefully the following article will explain the processes involved in cycling nitrogen.

Ammonia

The first stage in the Nitrogen Cycle begins as organic wastes start to decompose. These wastes originate from uneaten food, undetected fish corpses and excreta from the tank's inhabitants. The subsequent breakdown of the molecules in these wastes leads to the production of ammonia (if the pH of the water is neutral to alkaline) or ammonium ions (if the pH of the water is acidic). The bacteria responsible for the generation of ammonia are from the genera Bacillus, Bacterium or Micrococcus. The level of ammonia steadily increases as the tank ages, through the process of biochemical oxidation or ammonification. A peak of about 11mg/L is reached around day 7.

At this point in time, the bacteria responsible for the conversion of ammonia are numerous enough to reduce its concentration. The ammonia levels then drop abruptly and by day 11, its presence is virtually nil. If high levels are still detected beyond this time, or become apparent in an established tank several months later, a shortage of bacteria is indicated. There are two case scenarios to consider: increased ammonia is usually attributable to an excess of uneaten food or undiscovered corpses. A mass destruction of bacteria is normally a direct result of over zealous water changes, the use of medications, replacement or the washing of filter media in tap water or a substrate change.

A little scientific notation for the chemistry buffs.

As previously mentioned, the form of ammonia found in an aquarium is pH dependent. Free molecular ammonia (NH3), is a dissolved gas. It generally forms in tanks with a pH of 7 or greater (ie neutral to alkaline environments). This is the form that keepers of Malawi and Tanganyikan type biotopes need to be most aware of.

The toxicity of ammonia is increased further in tanks with warmer water. Under these conditions, ammonia is extremely toxic to fish and becomes fatal in a very short period of time. Biotopes that are acidic (ie have a pH less than 7), ammonia combines with a water molecule to form ammonium hydroxide (NH4OH). In this form, ammonia is approximately 100 times less dangerous to fish. Therefore, African cichlids are at a higher risk than their South American counterparts due to the highly mineralised and alkaline environment. The presence of free ammonia is often indicated by a darkening of fish colour, skin damage and gill irritation (observed through fish gasping for air at the surface). It becomes fatal above a concentration of 0.02mg/L.

How do I Rectify this Situation?

Partial water changes will reduce the concentration of ammonia in the tank, whilst the addition of new bacteria will help re-populate the colony. Products such as Cycle, Environ-8, Nitrovec etc. are useful. An ammonia test kit, either colorimetric (colour scale) or electronic, is an essential addition to the home aquarium.

Nitrite

The second stage of the Nitrogen cycle involves the activities of bacteria from the Nitrosomonas or Nitrosococcus genera. Nitrites (NO_2), play an integral part in the decomposition cycle of nitrogenous waste in the aquarium. They are produced when the afore-mentioned bacteria decompose the ammoniacal derivatives (NH_3 and NH_4). From around day 6 to day 17, they convert the ammonia into nitrites. At around day 24, the nitrites begin to decline until such time that they are undetectable.

Although not quite as toxic as NH_3 and NH_4, nitrites are harmful to fish regardless of the pH. Therefore, fish should not be introduced into an aquarium that shows its presence. Nitrous bacteria are practically non-existent in new aquaria. To start the process, you can introduce microorganisms that are readily sold at your LFS (Cycle, Environ-8, Nitrovec and so on). If you already have an established tank, you can call upon that reservoir of bacteria. Simply through the transfer of some filter material, aquarium decor or a handful of gravel, you can add some of the nitrous bacteria to your new tank. The transfer of water does not have the same beneficial effect as only a small amount of bacteria is suspended in the water column.

Nitrite levels should be checked regularly through the use of a colorimetric or electronic test kit. The slightest trace of nitrite can prove fatal and even when the levels register as zero, care should be taken to introduce fish gradually. Established aquariums are stable and all efforts to maintain the equilibrium must be observed. This includes not harming the bacteria through over zealous cleaning of the filter material. Do not use detergents or chlorinated water in the cleaning process and water changes should not involve replacing too large a volume of water. Fortnightly replacement of around 25% water is far more preferable to larger changes at less frequent intervals. Note that some medications (particularly antibiotics) can also destroy valuable bacteria. Isolating and medicating fish in a smaller tank is a better method of treatment.

Unless action is taken immediately, the fish's health will suffer drastically. The nitrite level should not exceed 0.01mg/L. Nitrites combine with hemoglobin (the oxygen carrying component in blood) and form methaemoglobin, thus preventing the transport of oxygen. The fish turns dark brown in colour and death ensues quite quickly. The presence of nitrites in the water is indicative of poor water quality, a situation that will aggravate quite a number of diseases.

How do I Rectify this Situation?

A lack of balance can arise when a weakness in the bacteria or a rise in nitrates (through oxygen depletion) occurs. This situation can be rectified through a water change and the introduction of supplementary bacteria.

Nitrate

From about the 5th day after filling an aquarium, nitrates begin to form, although it is only after about day 25 that the different populations that bring about the Nitrogen Cycle are in large enough numbers for all stages of the cycle to take place simultaneously. The intermediate toxic compounds, ammonia and nitrite, should be no longer detectable. The bacteria responsible for this activity are from the genus Nictrobacter.

This is the stage at which fish can be successfully added to the aquarium. It should be noted, however, that even though plants assimilate the nitrates originating from the breakdown of organic wastes, they are rarely in sufficient numbers to absorb all that is produced.

This fact becomes more apparent in many African type biotopes that are devoid of plants altogether. Over-populated, closed environments become subject to a steady build up of nitrates. An unattended tank can amass concentrations in excess of 1g/L over a

1 year interval (as opposed to a natural environment where the concentration rarely exceeds 5mg/L due to permanently changing water).

Excessive concentrations of nitrates are often termed 'silent killers'. This is because in small doses, these compounds are not deemed toxic. An accumulation that leads to concentrations in excess of 80-100mg/L will prove fatal to your fish. Only regular water changes will reduce their concentration. Certain species are more susceptible to nitrate poisoning than others. Amounts in excess of 50mg/L are often signaled by the explosive growth of algae. In a planted tank, preferred water plants are compromised by the algae that feed on the nitrogenous compounds. Affected fish will display signs of stress, hence making them more vulnerable to opportunistic diseases. Poor water quality can provoke such conditions as 'pop-eye' and fin rot. Care should be taken to keep nitrates under 15mg/L.

Additional latent risks can lead to dramatic consequences. For example, if the environment is oxygen depleted due to poor maintenance or a sudden rise in temperature, favourable conditions for the bacteria from the genus Bacillus develop. This, in turn, reverses the Nitrogen Cycle through denitrification hence producing ammonia and nitrite. If this is allowed to occur, the entire fish population can be decimated within a few hours.

How do I Rectify this Situation?

Commercial denitrifiers work on the principle of external denitrification. The nitrate charged water circulates through an anaerobic rich environment which breaks down the nitrates into molecular nitrogen. The provision of 'food' (for the bacterial colonies) is the only constraint and manufacturers have developed specially adapted foods in a number of different forms.

However, there is no substitute for regular water changes and testing for nitrate concentrations.

Nitrogen Cycle Summary

Ammonia is extremely toxic to fish. It causes swelling to the gills, leading to asphyxia and disruptive osmoregulatory functions. Nitrites become fatal by acting on the hemoglobin in the blood. Nitrates, however, are only toxic in high concentrations (between 50 and 300mg/L depending on the species).

Therefore, only when ammonia and nitrites have become undetectable, should fish be introduced into a newly installed tank.

This usually occurs around week 3-4. Various test kits are available to monitor the levels of ammonia, nitrites and nitrates-it is vital to use them regularly in a newly set-up tank. Those of us who are impatient can help to speed up the natural processes that promote the nitrogen cycle by introducing gravel or filter media from an established aquarium.

Where do Bacteria Live?

Bacteria are one of the most abundant organisms on the planet and there is some evidence that some of them may have even existed on Mars in the distant past. These tiny single-celled organisms are able to survive in many different habitats from deep under the sea to the Arctic, to warm water vents near volcanoes.

The soil on many parts of the Earth's surface are teeming with them, as are the surface of plants, animals and humans. Bacteria also grow very well inside the digestive system of animals and there are millions of them in the human intestine.

Strange Environments

Some of the strangest environments that support bacterial life include places that have extremes of temperature. The bacteria found here are also strange, much different to the bacteria found living in and around people. Most bacteria use carbon-based sugars as their substrates, whether they are aerobic bacteria that require oxygen, or anaerobic bacteria that can live without oxygen. Species of bacteria that live in the Arctic permafrost have completely different metabolic systems and use methane as their main energy source. Those that live in deep-sea vents, where no light is able to penetrate, use hydrogen sulphide.

Bacteria in Soil

There can be as many as three million bacteria in a gram of soil. Bacteria play a very useful role in soil, where they act as decomposers that recycle the organic matter from dead plants and animals. Many different types of bacteria manage the way that nitrogen, the major element present in proteins, recycles in the environment.

Some soil bacteria are able to fix nitrogen from the air, converting it to compounds that can be used by plants growing in the same soil. Others work on the decaying proteins from dead matter, converting it into nitrates that can be reused directly by nearby plants. Yet others carry out the opposite process, converting nitrogen-containing compounds from decaying plants and animals into nitrogen gas, which then re-enters the air. This recycling activity is vital in all habitats and is described as the nitrogen cycle.

Bacteria in Plants

Some nitrogen fixing bacteria actually live in the roots of plants, fixing nitrogen from the air and incorporating it into nitrogen compounds that can be used to build proteins for plant growth. Species that are part of the Rhizobium genus live in root nodules in plants such as peas and beans, which are known as legumes. The bacteria in the root nodules give the legumes the capacity to make large amounts of protein – which is why peas and beans are such a protein-rich food source.

Bacteria in and on Animals

The skin of animals and humans teems with living bacteria of many different species. Many of these do good work, crowding out and fighting off bacteria that cause disease. For this reason, they are often called friendly bacteria. All of the body openings also contain large populations of bacteria, particularly the mouth. These are good for the host but animal bites, even if they cause small puncture wounds, can be very dangerous because of all the bacteria that are introduced into the tissues of the animal or person who has been bitten.

The digestive system of animals is also a common place to find bacteria. Different animals have different types of bacteria – those present in a ruminant animal such as a cow secrete enzymes that are able to digest cellulose. This interaction enables the cow to digest large amounts of grass in the different chambers of its stomach.

In humans and other mammals, bacteria living in the gut help with digestion and synthesis useful compounds that cannot be made by the body systems. For example, humans are unable to make the vital vitamin B12; but it is synthesised in the intestines by specialised bacterial species that exist there as helpful gut bacteria.

Bacteria all Around us

Bacteria are all around us, which means that we do have to take care, even though most of them are harmless. Pathogenic bacteria can pass from person to person, and they can also survive in our homes and in the environment. Many bacterial diseases spread via food and water. Devastating epidemics of cholera were common in England in the 19th century, caused by sewage contamination of drinking water. Such epidemics still occur in developing countries, or in places that suffer a natural disaster such as a heavy flood or a hurricane. People who are infected with bacteria that cause food poisoning can also pass their infection on when they prepare food for others, making it sensible to take simple food hygiene precautions.

Bacteria and Antibiotics

Dealing with Antibiotic Resistance

Antibiotic resistance developed within a few years of the first antibiotic coming into widespread use in World War II. It is now a global problem. People all over the world, including in the industrialised countries where there are stringent regulations and highly trained medical professionals, continue to misuse antibiotics. Several key strategies will be required to prevent further antibiotic resistance developing and to stop the spread of strains that are already antibiotic resistant.

Reduce Antibiotic use in Animals

Around 87 per cent of the antibiotics used in veterinary medicine currently go to farm animals. This creates opportunities for bacteria present in farm animals to become resistant to antibiotics, and for these bacteria to pass on their antibiotic resistance genes to some of the bacteria that infect humans. It is also a fear that antibiotic residues in food will provide low levels of antibiotics that could further generate antibiotic resistance in human bacterial pathogens. Using antibiotics to prevent infections needs to stop, and treatment with antibiotics should be reserved for serious infections.

In the USA, an unusual loophole relating to animal treatment with antibiotics was reported in 2008. Pet shops in the USA that sell fish are also allowed to sell antibiotics to treat fish diseases. These are human antibiotics and can be bought from pet shops without a prescription. Being able to buy broad spectrum antibiotics freely in pet shops encourages people to buy them for their own use and is a highly dangerous practice that should be stopped through legislation.

Reduce Antibiotic use in Dentistry

Many dentists prescribe antibiotics unnecessarily and this has contributed to the antibiotic resistance seen in bacteria that causes tooth abscesses. Twenty years ago, only 5 per cent of all dental abscesses were due to antibiotic resistant bacteria but by the mid 1990s it was over 50 per cent.

The figure today is probably even higher but no official data has yet been made available. Dentists should ensure that they only prescribe antibiotics when necessary and should encourage patients to complete the full course.

Reduce Antibiotic use in People

There is no doubt that antibiotics have saved many lives since 1940 but they have also been used millions of times when they were not needed. Recent studies have shown that GPs in the UK frequently feel under pressure to prescribe antibiotics, even when they know that it is inappropriate. There is some evidence that they put a false diagnosis in notes so that their prescription will not be questioned. Although it is the responsibility of a GP to prescribe antibiotics sensibly, it is also up to the people who consult GPs to be aware of the problem of antibiotic resistance. Arguing heatedly with a GP when he or she advises that antibiotics are not indicated should be avoided.

Educate The Public

Part of the problem that arises in the GP's surgery stems from the great misunderstandings that exist in the mind of the general public. Very few people understand exactly what antibiotic resistance is, or the difference between bacteria and viruses. Many people assume that a common cold should be treated by antibiotics, not realising that colds are caused by viruses not bacteria. Governments around the world need to take steps to ensure that the general population is better educated about antibiotics. There are some projects to teach school children about the sensible and safe use of antibiotics but targeted campaigns aimed at older people would also be extremely valuable.

Abolish the Internet Trade

It is extremely easy to buy antibiotics over the Internet from countries that do not require a prescription for their sale. This is a great temptation to people in countries where prescriptions are required, but is another very dangerous practice. Not only can using antibiotics incorrectly generate resistance, there are no guarantees of the quality of the drug supplied. It may be a counterfeit product, with no active ingredient. Worse still, it could contain contaminants and toxins and cause illness or even death.

Set up Global Controls

The next obvious step from abolishing Internet sales would be to put international pressure on countries who still do allow antibiotics to be bought without prescription. Better control of antibiotic use needs to be implemented worldwide for the battle against antibiotic resistance to even have a chance of being won.

How Antibiotics Work

Antibiotics halt bacterial growth and cure infection through two main mechanisms. Antibiotics are either bactericidal or bacteriostatic. Bactericidal antibiotics kill the bacteria causing the infection through direct action, usually by causing the cells to split open, or lyse. Bacteriostatic antibiotics act on the internal workings of the bacterial cell to stop it dividing and so slow down the advance of the infection. A bacterial population that divides more slowly, or that cannot divide at all is much more easily dealt with by the body's immune system.

Bactericidal Antibiotics

Many bactericidal antibiotics work by altering the biochemical pathway through which bacteria make the cell wall. As the antibiotic is taken into the cell, it stops the biochemical machinery of the cell producing or attaching one the major components of cell wall structure.

The cell wall produced is thinner than usual. As the cell divides, the two daughter cells then also have weaker cell walls and they cannot strengthen them because they are also prevented from making all of the necessary components. As they try to divide subsequently, the cell walls of these daughter bacteria fail. Lysis of the cell follows and the bacterium dies. Penicillin antibiotics work in this way, as do the cephalosporins.

Other antibiotics in the aminoglycoside class are also bactericidal in some infections (they are bacteriostatic in others). They bind to part of an intracellular structure-the ribosome. This usually assembles amino acids together to form complete proteins. When the antibiotic is bound to the ribosome, it cannot make proteins efficiently, and fewer proteins, or proteins that contain mistakes, are made. Vital proteins that are required by the bacterium are therefore in short supply and the cell dies.

The quinolones disable bacterial enzymes that normally replicate bacterial DNA – making it impossible for the bacterium to divide. This happens quickly, and the affected bacteria die within a few hours. Quinolone antibiotics enter human cells very easily, so are useful for treating bacteria that penetrate and invade cells. The antibiotic has no toxic effects on the human cell itself because its own enzymes for copying DNA are completely different.

Narrow Spectrum and Broad Spectrum

Antibiotics are classed as bactericidal or bacteriostatic according whether they kill bacterial cells directly or indirectly. They are also divided into classes such as cephalosporins or macrolides depending in their chemical structure and action. All antibiotics can also be described as either narrow spectrum or broad spectrum. Those with a narrow spectrum of action can kill only a small number of species of bacteria, maybe even just one. Broad spectrum antibiotics are active against a wide range of bacterial species.

Narrow spectrum antibiotics tend to be very specific and act on a molecule in the metabolism of one particular type of bacteria that is special to that species. Broad spectrum antibiotics act on structures or processes that are common to many different bacteria, such as the components of the cell wall.

Choosing which Antibiotic

Doctors are sometimes faced with difficult choices. So many different bacteria cause infections, and there are many different antibiotics to choose from. The problem of antibiotic resistance is always an important consideration. Widespread use of broad spectrum antibiotics is one factor that is thought to have helped the spread of antibiotic resistance and doctors aim to prescribe antibiotics sensibly so that antibiotic resistance does not increase further.

The ideal treatment involves an antibiotic that is effective against the organism causing the infection and few others. Unfortunately, although taking samples of infected material from each patient and culturing the bacteria to make a positive identification would make this possible, this would not be a cost-effective practice because of the huge numbers of patients who go to their GP with an infection.

For most doctors, the choice ultimately depends on the seriousness of the infection, the antibiotics that are readily available, the most cost effective treatment, and the drug that will cause the fewest side effects.

How do Antibiotics Affect Friendly Gut Bacteria?

Antibiotics are useful drugs that can treat bacterial infections very successfully, despite the antibiotic resistance that has developed. Taking antibiotics does carry some risk of side effects, and it is best to take antibiotics only when an infection is severe enough to cause pain and distress, or is life threatening or a danger to others.

One of the main problems with any antibiotic is, while the drug is specific for bacterial cells, it cannot be made specific to only harmful bacteria. It is not possible to develop an antibiotic that does not kill normal friendly bacteria in the body as well as the pathogen causing the infection. Friendly bacteria grow in their millions on the skin, in openings to the body such as the mouth, nose and vagina, and in the intestines of the digestive system. They carry out important functions at all these sites but their primary role is to crowd out potential pathogens, preventing them from setting up an infection.

These potential pathogens include fungi such as Candida, as well as bacteria such as pathogenic strains of Staphylococcus aureus. When antibiotics destroy friendly bacteria, these potential pathogens can seize the opportunity and can grow much more easily as they have no competition. This is one of the reasons why antibiotic therapy often results in thrush, a fungal infection caused by Candida.

Effect on the Digestive System

Antibiotics are usually taken as a pill and swallowed. The capsule dissolves in the small intestine and the active ingredient passes across the intestinal wall and into the bloodstream to be carried round the body. The initial contact between the drug and the digestive system can cause some of the common side effects of nausea and vomiting. Any residual drug passes into the bacteria-rich environment of the large intestine and enhances the killing effect of the antibiotic circulating in the blood stream. The effect of this is to kill large numbers of friendly bacteria in the gut.

The Impact of Destroying Friendly Bacteria

The bacteria that fall victim to the antibiotic include bacteria that help keep other species under control and also bacteria that contribute to the digestion of food. Some bacteria in the human gut make vitamins such as vitamin K and some of the B vitamins. If they are destroyed, the body does not receive these vitamins and, if antibiotic therapy continues over an extended period, vitamin deficiency can result. Overgrowth of pathogenic species of bacteria can also occur. Bacteria such as Clostridium difficile are normally present in the large intestine, but their populations are kept to a minimum size by friendly bacteria.

After antibiotic therapy, the C difficile can grow rapidly as their usual constraints have disappeared. The bacteria produce toxins that irritate and kill the cells that line the large intestine, causing severe diarrhoea and internal bleeding. This infection happens frequently in elderly people whose immune systems are weakened by underlying illness, and C difficile infections have become very common in hospitals and nursing homes in the last 10 years.

Recently published research on patients treated with ciprofloxacin for infections of the urinary tract has actually measured the impact of antibiotic therapy on the proportions and numbers of friendly gut bacteria. Before treatment, the scientists detected around 5500 different strains and species of bacteria in the faeces of volunteers in the study. They were then given a typical course of the antibiotic and their gut bacteria were assessed again. About a third of the different types of bacteria had gone. Further monitoring showed that the bacteria did not return to their normal levels for about a month, suggesting that the health of the person treated could be affected long after the treatment had been completed.

Replenishing Friendly Bacteria

If you have to take a course of antibiotics, some doctors recommend that you eat plenty of natural yoghurt, containing live Lactobacillus bacteria during your treatment. These bacteria help to replenish the stocks of friendly bacteria in the large intestine and may lessen some of the effects of the treatment. Probiotic therapies are being trailed at the moment to see if they could be used to prevent some of the cases of C difficile in the elderly. Until they are proven, however, eating foods naturally high in friendly bacteria seems to be sensible and carries few risks.

Is it Possible to Develop new Antibiotics?

It is possible, yes, but at the current time, new antibiotics are being developed more slowly than at any time in their history. The pace of discovery was high in the 1940s and 1950s and efforts were renewed in the 1990s when MRSA became a problem worldwide.

In the late 1990s, Linezolid, a new class of antibiotics was developed and released for treating MRSA but resistance to it appeared just a few years later. In 2008, a company in the UK was starting to trial a new class of antibiotic that shows promise against MRSA, including vancomycin resistant strains. If successful, the new drugs could become available in the next two to three years. Although smaller biotechnology companies are still searching for new antibiotics, the larger pharmaceutical companies have moved away from this type of work. In the economic downturn that is affecting the world towards the end of the last decade of the 21st century, even the pharmaceutical giants are seeking to develop more profitable drugs to stop their business going into decline. Antibiotics, because they are taken in short courses, are much less profitable than drugs for chronic diseases, which could be taken for years by just one patient.

Where Might New Antibiotics Come From?

One possibility is that new antibiotics might be extracted from plants. Bacteria can infect plants as well as animals and plants produce their own chemicals to keep bacteria under control. Many such substances are known to exist in plants, but few have been studied or exploited as potential antibiotics. The rainforest plants, in particular, are considered an excellent source of plant molecules that could have antimicrobial activity.

Some of the antibiotics we currently use were extracted from bacteria themselves. A class of bacterium that also produces molecules with killing activity against species of bacteria that are pathogenic in humans are the Archaea bacteria. These include the bacteria that live in hot springs and that metabolise hydrogen sulphide. A study of just one hot spring in the USA showed that this single site had 48 different species of Archaea bacteria and some experts think this represents great untapped potential as a source of novel antibiotics.

Targeting RNA Polymerase

In the 1980s, a new class of antibiotics was discovered that act by disrupting the activity of a bacterial enzyme that sets off the production of RNA from the DNA template. This RNA template is used by the cell to make vital proteins. Myxopyronin has been shown to act in this way and to use this mechanism to prevent bacteria from making proteins, so killing them. Although some preclinical work has been done on myxopyronin, it has not yet been trailed in humans so is still several years away from being made available.

RAMP Controversy

Within the last few years, a new form of antibiotic called RAMP – short of ribosomally synthesised antimicrobial peptides – has been in development. These short stretches of protein molecules fatally interfere with the cell membrane of bacteria. They were thought to hold great promise but then research revealed that bacteria could become resistant to them. Since similar short peptides are used by the immune system of the body to fight infections, it is feared that resistance to RAMP would generate bacteria that would be resistant to the human immune system. These virtually indestructible bacteria would be disastrous and the RAMP drugs will probably never be used.

Phages as Antibiotics

Phages are not a new idea; Russian scientists have used them to treat bacterial infections for 60 years. Phages, short for bacteriophages, are viruses that infect bacteria. Lytic phages cause the bacteria they infect to split open, killing them, so have great potential as a biological antibiotic. There is currently much new interest in phages as they would be effective against the multiply drug resistant strains of Staphylococcus aureus that are currently such a problem in hospitals.

Major Classes of Antibiotics

The first antibiotics developed were penicillin and streptomycin and these are both still in use today, despite the development of antibiotic resistance during the last 70 years. Penicillin is now in a class of antibiotics known as the penicillins, which also includes the more recently developed antibiotics ampicillin, flucloxacillin and methicillin. These antibiotics all prevent bacteria making strong cell walls, leading to lysis and death of the bacteria.

Streptomycin, which was also developed in the 1940s, is now in the aminoglycoside class of antibiotics, which also has gentamycin, kanamycin and netromycin as members. All of these antibiotics bind to the smaller subunit of the bacterial ribosome, preventing the bacterium making new proteins. They work well to treat infections by gram negative bacteria such as Pseudomonas aeruginosa, Escherichia coli (E. coli) and Klebsiella bacteria that cause pneumonia. The aminoglycosides do not work against bacteria that grow well in the absence of oxygen.

The Cephalosporins

There are a great many antibiotics in this class. The earliest cephalosporin was isolated by an Italian scientist in 1948, from bacteria growing in a sewer on the island of Sardinia. The compound that was extract was found to be active against Salmonella typhi, the pathogen that was responsible for typhoid fever. The cephalosporins work in the same way as penicillins – they prevent proper production of a bacterial cell wall.

The earliest cephalosporins were active against just a few gram negative bacteria but chemists worked on the molecule, creating second, third, fourth and fifth generation cephalosporin antibiotics.

They generally increased their spectrum of activity, and the most recent ones can be used to treat infections caused by gram positive as well as gram negative bacteria.

The Tetracyclins

The tetracyclins also act on the bacterial ribosome to prevent the bacterial cell making vital proteins. As well as the drug tetracycline, the group also includes doxycycline and oxytetracycline. These antibiotics are not generally used for common bacterial infections apart from severe acne. They are used to treat infections by Mycoplasma species (tuberculosis and leprosy), Rickettsial species, Chlamydia species and the bacteria that cause Lyme disease, syphilis, anthrax and plague. Tetracyclins have quite significant side effects in addition to the usual digestive disturbances. They increase sensitivity of the skin to sunlight, they can affect the enamel on teeth, causing browning, and they can slow down bone growth

The Macrolides

Macrolide antibiotics are generally used to treat infections caused by streptococcal bacteria and respiratory infections generally. They are also active against syphilis, Lyme disease and leprosy and tuberculosis. Erythromycin is probably the most well known antibiotic in this class, which also includes clarithromycin and telithromycin, which is often used to treat pneumonia. The macrolides can have a toxic effect on the liver, causing liver toxicity and jaundice and so should not be given to people who have an underlying liver problem such as an infection with hepatitis C.

The Glycopeptides

Glygopeptide antibiotics such as vancomycin, telavancin and decaplanin prevent the adding of new units to the bacterial cell wall. Although they are similar in action to the penicillins and the aminoglycosides, the glycopeptides are extremely toxic and are used in few patients. They are reserved for people whose infections will not respond to other antibiotics because of antibiotic resistance. Until the development of newer antibiotics, glycopeptides were the standard treatment for MRSA.

The Quinolones

Quinolone antibiotics interfere with the bacterial enzymes bacterial DNA gyrase or the bacterial topoisomerase enzyme that help bacterial DNA to replicate during cell division. Commonly used quinolones include ciprofloxacin, lecofloxacin and norfloxacin. They are particularly useful antibiotics for treating infections of the urinary tract, pneumonia that is acquired in the community (as opposed to hospitals) and diarrhoea caused by pathogenic bacteria such as Shigella. The quinolones cause very few side effects.

The Newest Antibiotics

When MRSA first started to cause hospital epidemics in the 1990s, the search for new antibiotics to treat these resistant bacterial infections became intense. A new class of antibiotics, the oxazolidinones was developed, of which there is currently only one available drug, linezolid. This is active against MRSA and is reserved as the last line of treatment, and kept exclusively for this purpose. It is hoped that this will prevent or at least slow the development of bacterial resistance to it. Daptomycin, an antibiotic originally developed in the 1980s and then put on hold because of its toxic side effects, has been revived since the 1990s, again to treat MRSA. It has a unique mechanism of action and works by destroying the cell membrane of bacterial cells. It is active against gram positive bacteria generally but, like linezolid, is kept as a restricted treatment for infections caused by antibiotic resistance organisms.

What are Antibiotic Resistant Bacteria?

Antibiotic resistant bacteria are strains of bacteria that can carry on growing and dividing normally in the presence of an antibiotic that should kill them, or severely inhibit their growth. Antibiotic resistance started to develop shortly after penicillin, the first antibiotic was introduced in the early 1940s. Bacteria divide very quickly, as fast as once every 20 minutes, and so their rate of mutation can be exceptional.

When they are exposed to an environment that contains high levels of an antibiotic, most bacteria die. However, a mutation in just one single bacterium that enables it to withstand the effects of that antibiotic can lead to a whole population of antibiotic resistant bacteria. The gene that confers resistance can then be passed between bacteria during conjugation. This is a process in which bacteria link up with narrow communication tubes and pass DNA between them through the tubes. It is a common mechanism by which antibiotic resistance spreads within and between species of bacteria. It explains why antibiotic resistance has spread so quickly and efficiently throughout the world.

Enzyme Production and Drug Inactivation

Several classes of antibiotics, including the penicillins and the cephalosporins have a chemical structure called a beta lactam ring. This ring structure must be intact if the antibiotic is to prevent cell wall synthesis in the target bacteria. Bacteria can become resistant to penicillins and cephalosporins by gaining a gene that produces an enzyme capable of destroying the structure of the antibiotic. The enzyme is often called penicillinase but its modern chemical name is beta lactamase. Over 200 different forms of beta lactamase have been described, all of which have different chemical structures but all of which carry out the same function.

Bacteria that are able to resist the killing effects of chloramphenicol are able to destroy the antibiotic molecule using an acetyl transferase, an enzyme that removes all the acetyl groups and breaks the antibiotic apart. Other bacteria become resistant to beta lactam antibiotics by producing proteins that bind to the beta lactam ring, so that it cannot prevent cell wall synthesis. The binding protein in this case acts as a physical blocker rather than an enzyme that destroys the antibiotic.

Preventing Antibiotic Entry and Action

Some bacteria also change the proteins on their surface so that there are no entry points for the antibiotic. Methicillin-resistant Staphylococcus aureus (MRSA) has an altered penicillin binding protein that prevents any of the penicillin-like antibiotics gaining entry to the cell.

Other antibiotic resistant bacteria have altered target molecules inside their bacterial cell. Bacteria that become resistant to trimethoprim do so because the target enzyme of the antibiotic is slightly different to the enzyme in wild type bacteria and the antibiotic cannot recognise it.

Pumping out Mechanisms

Several species of bacteria have developed a molecular pump that moves the antibiotic molecules out of the bacterial cell faster than they can get in. E. coli and Pseudomonas aeruginosa are known to have multi-drug systems that move out many different antibiotics and drug molecules, preventing them from having any effect on the bacterium. Resistance to the tetracyclin class of antibiotics is now widespread in Enterobacteria and is due to a protein transport pump that uses energy to continually rid the cell of the drug.

Bypass Systems

Antibiotics that act on a particular metabolic pathway, in which one chemical is produced from other in a series of small steps, can stop working if the bacterium finds a way to make the product via a different route. Sulphonamides work by preventing bacteria producing a vital precursor molecule for the synthesis of DNA. Sulphonamide resistant bacteria do not need this precursor, and use a readily available alternative. However, the most famous example of a bypass system is that seen in MRSA, which produces an alternative to the normal penicillin binding protein. This binds the antibiotics, but the binding event has no effect on the function or survival of the bacterium.

New Resistance is Developing

Linezolid, one of the newest antibiotics that is now reserved for treating MRSA was introduced in the late 1990s and the first strains of linezolid resistant Staphylococcus aureus were detected in 2003, just a few years later. Whatever antibiotics are produced and used, bacteria will develop mechanisms that can resist their killing effects.

What is C Difficile?

C. difficile is Clostridium difficile, a member of the Clostridium family of bacteria. This family of dangerous human pathogens also includes Clostridium tetani, which causes tetanus, Clostridium botulinum, which causes botulism and Clostridium welchii and other species that cause gangrene. Until recently, Clostridium difficile was thought to be a harmless and friendly bacterium that was a normal inhabitant of the digestive system. It is commonly found in the large intestine of healthy people.

In the late 1970s, researchers realised that, under specific circumstances, C. difficile could cause disease. The victims were usually elderly patients in hospitals or nursing homes who had been treated with a course of antibiotics for an unrelated infection. During the course of treatment, they frequently developed severe diarrhoea and colitis and the main infective organism responsible was identified as C difficile. The course of antibiotics had wiped out the normal flora of the intestine, creating conditions that allowed C. difficile to change from being harmless to harmful.

Increase in Cases from the 1990s

Cases of antibiotic-associated diarrhoea due to C. difficile were probably increasing throughout the 1980s but the tools to diagnose the infection and characterise the strains involved were not developed until into the 1990s. This increase in diagnostic capacity is partly responsible for the increase in cases reported – from less than 1000 annually in the UK in 1990 to over 44 000 in 2004.

Further increases have been noted in the last four years, and the death toll has risen. In 2003, C difficile was mentioned as a contributory factor to death in 2000 death certificates in the UK, but in 2007, that figure had risen to 8324, a four-fold increase. One of the worst outbreaks in the UK occurred in 2006 at hospitals in the Leicester area. Although the cause of death was not confirmed in additional 29 cases, C. difficile definitely killed 49 people who were infected. Hospitals in Kent also reported a prolonged outbreak between 2004 and 2006, in which 90 people died.

Type 027

Since the bacterium C. difficile was first noted as a cause of post-antibiotic therapy diarrhoea, it has been researched more extensively. Over 100 types have now been described, according to the genes carried on the bacterial chromosome. One particularly dangerous strain, C. difficile Type 027 was first identified in the UK in 1999 and individual cases began to appear in the early years of the 21-century. However, in 2004 and 2005, there were several large outbreaks of C. difficile at the Exeter, Royal Devon and Stoke Mandeville hospitals. Typing of the isolates recovered from patients showed a very high level of Type 027. This type was also responsible for epidemic-style outbreaks in Canada and the USA.

Infection with Type 027 is more dangerous than other strains of C. difficile because it produces extra toxins that stimulate a higher rate of bacterial division and so cause more intestinal damage.

Dealing with C. Difficile

C. difficile is presenting a huge problem in hospitals all over the world as we approach the end of the first decade of the 21-century. Cases continue to increase and studies are showing that increased cleanliness and handwashing, whilst generally a good idea, does not really impact on C. difficile cases. The major cause is the use of antibiotics in hospitals – over half of all prescriptions are unnecessary and are due to out-dated standard practices, which are in need of urgent review.

For many patients, C. difficile infection subsides when the antibiotic course is completed, but the infection does recur a few weeks later in a high number of people. Treatment with metronidazole, an antibiotic specific for anaerobic bacterial will clear C. difficile, but subsequent reinfection is always a possibility. Various trials have looked at probiotic supplements taken with antibiotics to see if the introduction of friendly bacteria, mainly the Lactobacillus species, can prevent overgrowth of C. difficile in the intestine. Some positive results have been reported, but larger trials are needed.

It may also be possible to use a vaccine active against the toxins produced by C. difficile but, again, larger trials are needed before this could be used in hospital patients.

What is MRSA?

MRSA is an acronym used to describe a strain of Staphylococcus aureus that is resistant to the killing effect of the antibiotic methicillin. Methicillin-resistant Staphylococcus aureus is often called a 'superbug' because it is very difficult to treat and it causes a huge number of infections every year in hospitals all over the world.

Staphylococcus is a very common bacterium that is present in the upper respiratory tract and on the skin of many people. In most cases, it does not cause disease and is considered to be a commensal bacterium. However, the strains of Staph aureus that are carried by some people are pathogenic in others, particularly when the bacteria are introduced into a wound. It is possible for a healthcare worker in a hospital to carry Staph aureus without any ill effects but then to pass it onto patients who have recently had serious operations, causing a major post-operative infection.

How did Methicillin Resistance Arise?

All strains of Staphylococcus aureus were sensitive to penicillin when it was first manufactured and used to treat patients in the early 1940s. However, misuse of antibiotics quickly led to the development of resistance. Some strains obtained the gene for an enzyme that broke apart the penicillin molecule. By the end of the 1950s, virtually all strains of Staph aureus found in hospital patients were completely resistant to penicillin because of their production of penicillinase.

Researchers in the late 1950s and early 1960s tackled this situation by developing an alternative antibiotic – methicillin. This is in the same class of antibiotics as penicillin and it prevents bacteria from making new cell walls, just as penicillin does, but the molecule has a slightly different structure. This means that the penicillinase enzyme cannot attack it so bacteria that had developed resistance to penicillin were quickly killed by methicillin. This strategy worked well and strains of Staph aureus remained treatable by methicillin throughout the 1960s and 1970s. During the next decade, strains of methicillin resistant Staph aureus began to appear but the situation really became serious in the 1990s. Just as penicillin strains of the bacterium had done in the 1950s, strains of methicillin resistant Staph aureus gained the upper hand and caused widespread epidemics in hospitals throughout Europe and North America. The term MRSA technically means methicillin resistant strains of Staph aureus but is today used to describe Staph aureus that are resistant to all penicillin-like antibiotics.

What is MRSE?

Staphylococcus epidermidis bacteria, which are also common on the skin, have also gained methicillin resistance from their close relatives via plasmid transfer and are now a significant problem in intensive care units and surgical wards. These bacteria are now known as MRSE.

MRSA Epidemics

MRSA and MRSE have caused huge waves of infection in hospitals and are now considered a major risk for anyone entering hospital for an operation, particularly if they are elderly or weak from an underlying illness such as cancer. In recent years, hospitals have refocused on cleanliness and basic techniques such as handwashing, which help to reduce transmission of bacteria, and cases are currently declining.

Who is Affected?

In patients, MRSA can cause infection in any type of wound, but deep surgical wounds such as those necessary in a hip replacement operation or a heart or kidney transplant are a particular risk. Transplant patients are also treated with immunosuppresive drugs, which damp down their immune system so that they accept the graft better. However, these also reduce the body's ability to fend off bacterial infections, so transplant patients are among the most vulnerable to MRSA. MRSA can also infect the skin of frail and elderly patients, or those who stay in bed for long periods.

People with diabetes or varicose ulcers can easily become infected in breaks in the skin. The bacteria also take advantage of entry points into the body that are created by treatments such as injections and the introduction of IV lines, catheters and ventilation tubes. A local infection in an IV line can quickly become life threatening as the bacteria gain direct access to the bloodstream, causing septicaemia.

Even if this does not happen, the bacteria can track their way deep into the body but more slowly and can set up deep abscesses in the tissues and major organs. These are difficult to detect until the patient becomes extremely ill with a high fever, shock and a high white cell count in their blood.

Intubation of patients in intensive care units can introduce MRSA directly into the lungs, where it causes a highly fatal form of pneumonia. This illustrates the opportunistic nature of MRSA – this is one of the very few circumstances in which Staph aureus can cause a lung infection.

When were Antibiotics Discovered?

Sulphonamide drugs, which could be used to treat bacterial infections in people, were available from 1935 but the first true antibiotic did not come into widespread use until 1940. This was penicillin, which was actually discovered in 1928.

The discovery of penicillin by Alexander Fleming is one of the most famous stories in the history of science. It is an example of a discovery made by serendipity – in other words, by accident. Fleming happened to notice that a culture plate of bacteria that had been left lying around over a weekend had grown a mould. There was nothing unusual about that and Fleming was about to throw the plate away when he noticed that there was a clear area around the mould. He became intrigued when he realised that the mould was producing a substance that could kill the bacteria.

This led to further research and Fleming eventually extracted the active product from cultures of the mould. It was named penicillin after the Penicillium mould that had produced it. Although far from the purified penicillin we use today, this rough extract was still powerful enough to cure a few mice from a severe bacterial infection, suggesting that it might do the same in humans.

Chapter

4

Bacteria, Growth and Reproduction

Forensic scientists often culture and grow bacteria found at crime scenes or extracted from remains. This process is often necessary to achieve a large enough population of bacteria upon which tests can then be performed. An understanding of how bacteria grow, multiply, and change over time also helps explain many field or autopsy findings.

A population of bacteria in a liquid medium is referred to as a culture. In the laboratory, where growth conditions of temperature, light intensity, and nutrients can be made ideal for the bacteria, measurements of the number of living bacteria typically reveals four stages, or phases, of growth, with respect to time. Initially, the number of bacteria in the population is low. Often the bacteria are also adapting to the environment. This represents the lag phase of growth. Depending on the health of the bacteria, the lag phase may be short or long. The latter occurs if the bacteria are damaged or have just been recovered from deep-freeze storage.

After the lag phase, the numbers of living bacteria rapidly increases. Typically, the increase is exponential. That is, the population keeps doubling in number at the same rate. This is called the log or logarithmic phase of culture growth, and is the time when the bacteria are growing and dividing at their maximum speed. The explosive growth of bacteria cannot continue forever in the closed conditions of a flask of growth medium. Nutrients begin to become depleted, the amount of oxygen becomes reduced, the pH changes, and toxic waste products of metabolic activity begin to accumulate. The bacteria respond to these changes in a variety of ways to do with their structure and activity of genes. With respect to bacteria numbers, the increase in the population stops and the number of living bacteria plateaus. This plateau period is called the stationary phase. Here, the number of bacteria growing and dividing is equaled by the number of bacteria that are dying.

Finally, as conditions in the culture continue to deteriorate, the proportion of the population that is dying becomes dominant. The number of living bacteria declines sharply over time in what is called the death or decline phase. Bacteria growing as colonies on a solid growth medium also exhibit these growth phases in different regions of a colony. For example, the bacteria buried in the oldest part of the colony are often in the stationary or death phase, while the bacteria at the periphery of the colony are in the actively-dividing log phase of growth.

Culturing of bacteria is possible such that fresh growth medium can be added at rate equal to the rate at which culture is removed. The rate at which the bacteria grow is dependent on the rate of addition of the fresh medium. Bacteria can be tailored to grow relatively slow or fast and, if the set-up is carefully maintained, can be maintained for a long time. Bacterial growth requires the presence of environmental factors. For example, if a bacterium uses organic carbon for energy and structure (chemoheterotrophic bacteria), then sources of carbon are needed. Such sources include simple sugars (glucose and fructose are two examples). Nitrogen is needed to make amino acids, proteins, lipids and other components. Sulphur and phosphorus are also needed for the manufacture of bacterial components. Other elements, such as potassium, calcium, magnesium, iron, manganese, cobalt and zinc are necessary for the functioning of enzymes and other processes.

Bacterial growth is also often sensitive to temperature. Depending on the species, bacteria exhibit a usually limited range in temperatures in which they can grow and reproduce. For example, bacteria known as mesophiles prefer temperatures from 20°C–50°C (68°F–122°F). Outside this range growth and even survival is limited. Other factors, which vary depending on species, required for growth include oxygen level, pH, osmotic pressure, light, and moisture.

The events of growth and division that are apparent from measurement of the numbers of living bacteria are the manifestation of a number of molecular events. At the level of the individual bacteria, the process of growth and replication is known as binary division. Binary division occurs in stages. First, the parent bacterium grows and becomes larger. Next, the genetic material inside the bacterium uncoils from the normal helical configuration and replicates. The two copies of the genetic material migrate to either end of the bacterium. Then a cross-wall known as a septum is initiated almost precisely at the middle of the bacterium. The septum grows inward as a ring from the inner surface of the membrane. When the septum is complete, an inner wall has been formed, which divides the parent bacterium into two so-called daughter bacteria. This whole process represents the generation time.

Bacteria can exchange genetic material via conjugation. Genetic recombination between bacteria (or protists) occurs via a cytoplasmic bridge between the organisms. A primitive form of exchange of genetic material between bacteria involving plasmids also can occur. Plasmids are small, circular, extrachromosomal DNA molecules that are capable of replication and are known to be capable of transferring genes among bacteria. For example, resistance plasmids carry genes for resistance to antibiotics from one bacterium to another, while other plasmids carry genes that confer pathogenicity (the ability to cause disease). In addition, the transfer of genes via bacteriophages—viruses that specifically parasitize bacteria—also serves as a means of genetic recombination.

Bioengineering uses sophisticated techniques to purposely transfer DNA from one organism to another in order to give the second organism some new characteristic it did not have previously. For example, in a process called transformation, antibiotic susceptible bacteria that are induced to absorb manipulated plasmids placed in their environment can acquire resistance to that antibiotic substance due to the new genes they have incorporated. Similarly, in a process called transfection, specially constructed viruses are used to artificially inject bioengineered DNA into bacteria, giving infected cells some new characteristic. Evolution has driven both bacterial diversity and bacterial adaptation. Some alterations are reversible, disappearing when the particular pressure is lifted. Other alterations are maintained and can even be passed on to succeeding generations of bacteria.

Uses and Classifications of Bacteria

Bacteria are prokaryotes, that is, they are organisms that lack membranes surrounding their genetic material. Bacteria is reproduce mainly by simple, or binary, fission (splitting). Under certain condition, a mature organism splits apart to form two new organisms, it can multiply very fast and it can double its number every 20minutes. Budding is another way bacteria reproduce, it is a process in which a small out-growth from a single parent develops into a new, identical organism. There are also few bacteria that can reproduce sexually under certain conditions, this happen when two bacteria unite to form a zygote, which splits into new cells. There are some bacteria that can form a colonies big enough to be seen without using a microscope.

Uses of Bacteria

- A bacterium breaks down the organic fertilizer (decomposed vegetables and animal matter) into material that can be used by plants.
- Some species of soil bacteria convert nitrogen into nitrites, compounds that are readily absorbed by plants.
- Different commercial processes also need certain bacteria, like Anaerobic bacteria that ferment certain substances are used in the production of vinegar and some drugs, and in the aging process of cheeses.
- Some bacteria also produce waste products that are beneficial to humans, like Lactic acid, it is produced by intestinal bacteria and can promotes digestion in humans.
- Bacteria are also grown commercially and it is added to certain foods like yogurt and drinks.
- Bacteria are also used to chemically break down the tough, woody tissues of flax, jute, hemp, and coconut.
- Bacteria are also being used in modern sewage disposal known as Bioremediation, it is a process by which bacteria are added to water or soil to convert toxic pollutants, such as pesticides and oil, into harmless substances.

Bioremediation-is defined as any process that uses microorganism to return the environment altered by contaminants to its original condition. There is also a kind of bacteria that can cause somediseases, it is called pathogenic bacteria. Bacteria enter humans in different ways. Airborne bacteria enter through broken skin or by passing directly into the nose, throat, or lungs. They cause such diseases as diphtheria, pertussis (whooping cough), staph infections, strep throat, tetanus, and tuberculosis. Some bacteria enter the body means of contaminated food or and water. They can causes diseases like typhoid fever, dysentery, cholera, and botulism. Some strains of bacteria secrete toxins (poisons) into the bloodstream, which carries them to other parts of the body.

Different Uses and Classifications of Bacteria Continuation

Treatments/Prevention: Some bacterial diseases can be prevented by routine immunizations during childhood. Recommended natural remedy for bacterial infection are the following;

- Garlic-Thyme-natural and effective antibiotic. Strengthen the immune system
- Bee Propolis-another natural and strong antibiotic
- Vitamin C-a dosage of Vitamin C can help fight diseases caused by bacteria.
- Aloe Gel-loaded with amino acid and can strengthen the immune system
- Lots of water and fruit juices is important to clean the system

Classification of Bacteria based on shape:

- Bacilli (singular: Bacillus), rod-shaped bacteria; the most numerous of all types. They include coccobacilli and streptobacilli.
- Cocci-(singular: Coccus), spherical bacteria. This group is divided is divided into bacteria that occur in pairs, such as the staphylococci; in clusters, such as the streptococci.
- Spirilla-(singular: Spirillum), spiral-shaped bacteria; the least numerous type.
- Vibrios-S-shaped or comma-shaped bacteria.

Classification of Bacteria based on size, colour, and chemical composition.

- Order Actinomycetales-branching, filamentous bacteria characterized by fungus-like growth. They are important in soil fertilization and as a source of antibiotics.
- Order Beggiatoales-filamentous bacteria that move along a surface in a creeping or gliding motion. They use photosynthesis to manufacture food. Many oxidize sulphur compounds.
- Order Chlamydobacteriales-bacteria encased in a protective covering called a sheath. They grow in algae like clusters and use iron compounds for growth.

- Order Eubacteriales-rigid-cell bacteria that are rod-shaped or coccus-shaped. Most are not moving.
- Order Hyphomicrobiales-rigid cell bacteria that reproduce by budding.
- Order Myxobacteriales-flexible, rod shaped bacteria. They excrete slime and some produce appendages that are visible to the naked eye.
- Order Rickettsiales-small, rod shaped or coccus-shaped bacteria that are internal parasites in humans and other animals. Most are pathogenic.
- Order Spirochaetales-the spirochetes is a long thin, spiral shaped bacteria. Most are pathogenic.

Methods for DNA Isolation

Large Scale Double-stranded DNA Isolation

The method used for the isolation of large scale cosmid and plasmid DNA is an unpublished modification of an alkaline lysis procedure followed by equilibrium ultracentrifugation in cesium chloride-ethidium bromide gradients. Briefly, cells containing the desired plasmid or cosmid are harvested by centrifugation, incubated in a lysozyme buffer, and treated with alkaline detergent. Detergent solubilized proteins and membranes are precipitated with sodium acetate, and the lysate is cleared first by filtration of precipitate through cheesecloth and then by centrifugation. The DNA-containing supernatant is transferred to a new tube, and the plasmid or cosmid DNA is precipitated by the addition of polyethylene glycol and collected by centrifugation. The DNA pellet is resuspended in a buffer containing cesium chloride and ethidium bromide, which is loaded into polyallomer tubes and subjected to ultracentrifugation overnight. The ethidium bromide stained plasmid or cosmid DNA bands, equilibrated within the cesium chloride density gradient after ultracentrifugation, are visualized under long wave UV light and the lower band is removed with a 5 cc syringe. The intercalating ethidium bromide is separated from the DNA by loading the solution onto an equilibrated ion exchange column. The A260 containing fractions are pooled, diluted, and ethanol precipitated, and the final DNA pellet is resuspended in buffer and assayed by restriction digestion as detected on agarose gel electrophoresis.

During the course of this work several modifications to the above protocol were made. For example, initially cell growth times included three successive overnight incubations, beginning with the initial inoculation of 3 ml of antibiotic containing media with the plasmid or cosmid-containing bacterial colony, and then increasing the culture volume to 50 ml, and then to 4 l. However, it was observed that recombinant cosmid DNA isolated from cell cultures grown under these conditions, in contrast to recombinant plasmid DNA, was contaminated with deleted cosmid DNA molecules. However, these deletions are avoided by performing each of the three successive incubations for eight hours instead of overnight, although a slight yield loss accompanied the reduced growth times.

Recently, a diatomaceous earth-based method was used to isolate the plasmid or cosmid DNA from a cell lysate. The cell growth, lysis, and cleared lysate steps are performed as described above, but following DNA precipitation by polyethylene glycol, the DNA pellet is resuspended in RNase buffer and treated with RNase A and T1. Nuclease treatment is necessary to remove the RNA by digestion since RNA competes with the DNA for binding to the diatomaceous earth. After RNase treatment, the DNA containing supernatant is bound to diatomaceous earth in a chaotropic buffer of guanidine hydrochloride by incubation at room temperature. The DNA-associated diatomaceous earth then is collected by centrifugation, washed several times with ethanol buffer and acetone, dried, and then resuspended in buffer. The DNA is eluted during incubation at 65degC, and the DNA-containing supernatant is collected after centrifugation and separation of the diatomaceous earth particles. The DNA recovery is measured by taking absorbance readings at 260 nanometers. After concentration by ethanol precipitation, the DNA is assayed by restriction digestion.

Protocol:

1. Pick a colony of bacteria harbouring the plasmid or cosmid DNA of interest into a 12 X 75 mm Falcon tube containing 2 ml of LB media supplemented with the appropriate antibiotic (typically ampicillin at 100 ug/ml) and incubate at 37deg C 8-10 hours with shaking at 250 rpm. Transfer the culture to an Ehrlenmeyer flask containing 50 ml of similar media, and incubate further for 8-10 hours. Transfer 12.5 ml of the culture to each of 4 liters of similar media, and incubate for an additional 8-10 hours.
2. Harvest the cells by centrifugation at 7000 rpm for 20 minutes in 500 ml bottles in the RC5-B using the GS3 rotor. Resuspend the cell pellets in old media and transfer to two bottles, centrifuge as before, and decant the media. The cell pellets can be frozen at-70degC at this point.

3. Resuspend the cell pellets in a total of 70 ml of GET/Lysozyme solution (35 ml for each bottle) by gently teasing the pellet with a spatula and incubate for 10 minutes at room temperature. (Note: Do not vortex the lysate at any time because this may shear the chromosomal DNA).
4. Add a total of 140 ml of alkaline lysis solution (70 ml for each bottle), gently mix, and incubate for 5 minutes in an ice-water bath.
5. Add 105 ml of 3M NaOAc, pH 4.8 (52.5 ml for each bottle), cap tightly, gently mix by inverting the bottle a few times, and incubate in an ice-water bath for 30-60 minutes.
6. Clear the lysate of precipitated SDS, proteins, membranes, and chromosomal DNA by pouring through a double-layer of cheesecloth. Transfer the lysate into 250 ml centrifuge bottle, centrifuge at 10,000 rpm for 30 minutes at 4deg C in the RC5-B using the GSA rotor.

For cesium chloride-gradient purification:

7a. Pool the cleared supernatants into to a clean beaker, add one-fourth volume of 50% PEG/0.5 M NaCl, swirl to mix, and incubate in an ice-water bath for 1-2 hours.

8a. Collect the PEG-precipitated DNA by centrifugation in 250 ml bottles at 7000 rpm for 20 minutes at 4degC in the RC5-B using the GSA rotor.

9a. Dissolve the pellets in a combined total of 32 ml of 100:10 TE buffer, 5 ml of 5 mg/ml ethidium bromide, and 37 g cesium chloride (Var Lac Oid Chemical Co., Inc.) (final concentration of cesium chloride should be 1 g/ml).

10a. Transfer the sample into 35 ml polyallomer centrifuge tubes, remove air bubbles, seal with rubber stoppers, and crimp properly.

11a. Centrifuge at 60,000 rpm to 16-20 hours at 15-20degC in the Sorvall OTD-75B ultracentrifuge (DuPont) using the T-865 rotor.

12a. Visualize the ethidium bromide stained DNA under long-wave UV light, and remove the lower DNA band using a 5 cc syringe and a 25 gauge needle. (It may be helpful first to remove and discard the upper band).

13a. To remove the ethidium bromide, load the DNA sample onto an equilibrated 1.5 ml Dowex column, and collect 0.5 ml fractions. Equilibrate the Dowex AG resin (BioRad) by successive centrifugation, resuspension, and decanting with 1M NaOH, water, and then 1M Tris-HCl, pH 7.6 until the Dowex solution has a pH of 7.6.

14a. Pool fractions with an A260 of 1.00 or greater into 35 ml Corex glass tubes, add one volume of ddH2O, and ethanol precipitate by adding 2.5 volumes of cold 95% ethanol. Incubate at least 2 hours at-20degC, centrifuge at 10,000 rpm for 45 minutes in the RC5-B using the SS-34 rotor. Gently decant the supernatant, add 80% ethanol, centrifuge as before, decant, and dry the DNA pellet in a vacuum oven.

15a. Resuspend the DNA in 10:0.1 TE buffer.

For diatomaceous earth-based purification:

7b. Pool the supernatants from step 6 into 500 ml bottles and add DNase-free RNase A and RNase T1 such that the final concentration of RNase A is 40 ug/ml and RNase T1 is 40 U/ml. Incubate in a 37degC water bath for 30 minutes.

8b. Add an equal volume of isopropanol and precipitate at room temperature for 5 minutes. Centrifuge at 9,000 rpm for 30 minutes in the RC5-B using the GS3 rotor. Decant the supernatant and drain the DNA pellet.

9b. Resuspend each DNA pellet in 20 ml 10:1 TE buffer, and add 40 ml of de-fined diatomaceous earth in guanidine-HCl (100 mg/ml) to each bottle. Allow the DNA to bind at room temperature for 5 minutes with occasional mixing. Centrifuge at 9,000 for 10 minutes in the RC5-B using the GS3 rotor.

10b. Decant the supernatant, resuspend each pellet in 40 ml of diatomaceous earth-wash buffer, and centrifuge as above.

11b. Decant the supernatant, resuspend each pellet in 40 ml of acetone, and centrifuge as above.

12b. Decant the supernatant and dry the pellet in a vacuum oven.

13b. Resuspend the pellet in 20 ml of 10:1 TE buffer, and elute the bound DNA by incubation at 65degC for 10 minutes with intermittent mixing.

14b. Remove the diatomaceous earth by centrifugation at 9,000 rpm for 10 minutes in the RC5-B using the GS3 rotor. Repeat if necessary.

15b. Combine the DNA-containing supernatants and precipitate the DNA in 35 ml Corex glass tubes adding 2.5 volumes of cold 95% ethanol/acetate.

16b. Resuspend the dried DNA pellet in 2 ml of 10:0.1 TE buffer and assay for concentration by absorbance readings at 260 nm or by agarose gel electrophoresis.

Midiprep Double-stranded DNA Isolation

A midi-prep double-stranded DNA isolation has been developed to generate a sufficient amount of template DNA for several Sequenase[TM] catalyzed fluorescent terminator reactions. Here, one bacterial colony which harbored the plasmid of interest is picked into 3 ml of liquid media containing ampicillin and incubated in a 37degC shaker for 8-10 hours. At this time, the culture is transferred into 50 ml of ampicillin-containing media and incubated further for 10-12 hours.

After harvesting the cells by centrifugation, a diatomaceous earth-based alkaline-lysis purification method (19-22) is performed, similar to that discussed above for large scale DNA isolation. The purified DNA is crudely assayed for concentration and purity by agarose gel electrophoresis against known standards. The approximate yield of double-stranded DNA using this method is 1 ug of DNA per ml of cell culture. For a 50 ml cell culture, about 50 ug of DNA are recovered, and 5 ug are used typically in a Sequenase[TM] terminator reaction.

Note: This procedure is the method of choice for isolating double stranded plasmid-based templates for the Sequenase Dye-Labelled Terminator Sequencing Reactions.

Protocol:

1. Pick a colony of bacteria harbouring the plasmid DNA of interest into a 12 X 75 mm Falcon tube containing 3 ml of 2xTY media supplemented with the appropriate antibiotic (typically ampicillin at 100 ug/ml) and incubate at 37deg C 8-10 hours with shaking at 250 rpm. Transfer the culture to an Ehrlenmeyer flask containing 50 ml of similar media, and incubate further for 11-14 hours.
2. Harvest the cells by centrifugation at 3000 rpm for 5 minutes in 50 ml conical tubes in the Beckman GPR tabletop centrifuge and decant the supernatant. The cell pellets can be frozen at-70degC at this point.
3. Resuspend the cell pellets in 2 ml of GET/Lysozyme solution, add 4 ml of alkaline lysis solution, gently mix, and incubate for 5 minutes in an ice-water bath.

4. Add 4 ml of 3M NaOAc, pH 4.8, gently mix by swirling, and incubate in an ice-water bath for 30-60 minutes.
5. Clear the lysate of precipitated SDS, proteins, membranes, and chromosomal DNA by pouring through a double-layer of cheesecloth into a new 50 ml conical tube. Centrifuge at 3,000 rpm for 20 minutes at 4degC in the Beckman GPR tabletop centrifuge.
6. Decant the supernatant to a 50 ml polypropylene centrifuge tube, add 20 ul of a 20 mg/ml DNase-free RNase A and incubate in a 37degC water bath for 30 minutes.
7. Add 7 ml (equal volume) of de-fined diatomaceous earth in guanidine-HCl (20 mg/ml) and allow the DNA to bind at room temperature for 5 minutes with occasional mixing. Centrifuge at 3,000 for 5 minutes in the Beckman GPR tabletop centrifuge.
8. Decant the supernatant, resuspend in 7 ml of diatomaceous earth-wash buffer, and centrifuge as above.
9. Decant the supernatant, resuspend in 7 ml of acetone, and centrifuge as above.
10. Decant the supernatant and dry in a vacuum oven.
11. Resuspend the pellet in 0.6 ml of 10:1 TE buffer, and elute the bound DNA by incubation at 65degC for 10 minutes with intermittent mixing.
12. Remove the diatomaceous earth by centrifugation at 3,000 rpm for 5 minutes in the in the Beckman GPR tabletop centrifuge.
13. Transfer the supernatant to a 1.5 ml microcentrifuge tube and centrifuge at 12,000 rpm for 5 minutes in a microcentrifuge at room temperature. Transfer the supernatant to a new 1.5 ml microcentrifuge tube and ethanol precipitate.
14. Resuspend the dried DNA pellet in 40 ul of 10:0.1 TE buffer and assay for concentration by agarose gel electrophoresis.

Miniprep Double-stranded DNA Isolation

The standard method for the miniprep isolation of plasmid DNA includes the same general strategy as the large scale isolation. However, smaller aliquots of antibiotic containing liquid media inoculated with plasmid-containing cell colonies are incubated in a 37degC shaker for 12-16 hours. After collecting the plasmid containing cells by centrifugation, the cell pellet is resuspended in a hypotonic sucrose buffer. The cells are successively incubated with an RNase-lysis buffer, alkaline detergent, and sodium acetate.

The lysate is cleared of precipitated proteins and membranes by centrifugation, and the plasmid DNA is recovered from the supernatant by isopropanol precipitation. The DNA is crudely checked for concentration and purity using agarose gel electrophoresis against known standards. A typical yield for this method of DNA isolation is 10-15 ug of plasmid DNA from a 6 ml starting culture.

Since highly supercoiled DNA is desired for double-stranded DNA sequencing, a modification of this method employing diatomaceous earth sometimes is used for isolation of double-stranded templates for DNA sequencing with fluorescent primers. After removal of the precipitated proteins and membranes, the plasmid containing supernatant is incubated with diatomaceous earth and guanidine hydrochloride and this mixture is added into one of the twenty-four wells in the BioRad Gene Prep Manifold. The supernatant is removed by vacuum filtration over a nitrocellulose filter. The DNA-associated diatomaceous earth is washed to remove the guanidine hydrochloride with an ethanol buffer, and then dried by filtration. Elution buffer is added to the wells, and the DNA-containing solution then is separated from the diatomaceous earth particle by filtration into a collection tube. The collected DNA is concentrated by ethanol precipitation and crudely assayed for concentration and purity by agarose gel electrophoresis against known standards. The approximate yield of double-stranded DNA is 3-5 ug of DNA from 6 ml of starting culture.

Protocol:

1. Pick a colony of bacteria harbouring the plasmid DNA of interest into a 17 X 100 mm Falcon tube containing 6 ml of TB media supplemented with the appropriate antibiotic (typically ampicillin at 100 ug/ml) and incubate at 37deg C 16-18 hours with shaking at 250 rpm.
2. Harvest the cells by centrifugation at 3000 rpm for 5 minutes in the Beckman GPR tabletop centrifuge and decant the supernatant. The cell pellets can be frozen at-70degC at this point.
3. Resuspend the cell pellets in 0.2 ml of TE-RNase solution (50:10 TE buffer containing 40 ug/ml RNase A; some also add RNase T1 to a final concentration of 10 U/ul) by gentle vortexing, add 0.2 ml of alkaline lysis solution, gently mix, and incubate for 15 minutes at room temperature.
4. Add 0.2 ml of 3M NaOAc, pH 4.8, gently mix by swirling, transfer to 1.5 ml microcentrifuge tubes, and incubate in an ice-water bath for 15 minutes.

5. Clear the lysate of precipitated SDS, proteins, membranes, and chromosomal DNA by centrifugation at 12,000 rpm for 15 minutes in a microcentrifuge at 4deg C.
6. Transfer the supernatant to a fresh 1.5 ml microcentrifuge tube, incubate in an ice-water bath for 15 minutes, centrifuge as above for an additional 15 minutes and transfer the supernatant to a clean 1.5 ml tube.

For standard alkaline lysis purification:

7a. Precipitate the DNA by adding 1 ml of 95% ethanol, and resuspend the dried DNA pellet in 100-200 ul 10:0.1 TE buffer. Electrophorese an aliquot of the DNA sample on a 0.7% agarose gel to crudely determine the concentration and purity.

For diatomaceous earth-base purification:

7b. Add 1 ml of de-fined diatomaceous earth in guanidine-HCl (20 mg/ml) and allow the DNA to bind at room temperature for 5 minutes with occasional mixing. Meanwhile soak the Prep-A-Gene nitrocellulose membrane in isopropanol for at least 3 minutes, and assemble the Prep-A-Gene manifold as described in the manual.

8b. Turn on the vacuum pump and adjust the vacuum level to 8 in. Hg, let the membrane dry for 1 minute, and then release the vacuum.

9b. Pour the well mixed samples into the wells of the Prep-A-Gene manifold and filter through at 8 in. Hg until all the liquid is filtered through.

10b. Wash the samples four times with 250 ul of diatomaceous earth-wash buffer, using a repeat pipette, allowing all of the liquid to filter through between washes.

11b. Reduce the vacuum to 5 in. Hg before turning the vacuum off at the stopcock. Without unscrewing the black clamps, release the white clamps and place the collection rack with clean 1.5 ml screw-capped tubes into the manifold. Clamp the manifold with the white clamps, and apply 300 ul of 10:1 TE buffer heated to 65degC and pull the eluted DNA through at 5 in. Hg. After the liquid has filtered through, raise the vacuum to 10-12 in. Hg, and let the membrane dry for 1 minute.

12b. Turn off the vacuum at the stopcock and remove the collection rack containing the tubes. Ethanol precipitate the DNA and resuspend the dried DNA pellet in 30 ul of 10:0.1 TE buffer.

Large Scale M13RF Isolation

Double-stranded M13RF is isolated for use in M13 SmaI cut, dephosphorylated vector preparation, described below. The growth conditions of M13-infected bacterial cells appears convoluted, but result in a maximal amount of M13 RF molecules per cell. After the M13RF containing bacterial cells are harvested by centrifugation, the double-stranded molecules are isolated using the cesium chloride method for large scale plasmid isolation, as described above. This briefly entailed alkaline cell lysis, sodium acetate precipitation of detergent solubilized proteins and membranes, polyethylene glycol DNA precipitation, and extraction of ethidium bromide stained DNA from a cesium chloride gradient after ultracentrifugation.

After removal of the ethidium bromide on an ion-exchange column, the DNA containing fractions are detected by A260 measurement and pooled, and the DNA is concentrated by ethanol precipitation and assayed by restriction enzyme digestion and agarose gel electrophoresis.

Protocol:

1. Prepare an early log phase culture of JM101 by inoculating an Ehrlenmeyer flask containing 50 ml of 2xTY with a glycerol stock of JM101 and pre-incubating for 1 hour in a 37degC water bath, with no shaking. Pick a plaque representing the desired M13 clone into four 1.5 ml aliquot of early log phase JM101, and incubate according to the procedure displayed in Figure 1 to result in 4 liters of M13-infected bacteria.
2. Harvest the cells by centrifugation at 7000 rpm for 20 minutes in 500 ml bottles in the RC5-B using the GS3 rotor. Resuspend the cell pellets in fresh 2xTY media to remove contaminating extracellular phage and transfer to two bottles, centrifuge as before, and decant the media. The cell pellets can be frozen at-70degC at this point.
3. Resuspend the cell pellets in a total of 120 ml (30 ml for each bottle) of 1X STB buffer by gently teasing the pellet with a spatula. Add a total 24 ml of lysozyme solution (6 ml for each bottle), gently mix, and incubate for 5 minutes in an ice-water bath.
4. Add 48 ml of 50:2:10 TTE buffer (12 ml for each bottle) and 2 ml of RNase A (10 mg/ml) (0.5 ml for each bottle), gently mix, and incubate in an ice-water bath for 5 minutes.
5. Clear the lysate of precipitated SDS, proteins, membranes, and chromosomal DNA by pouring through a double-layer of cheesecloth. Transfer the lysate into 250 ml centrifuge bottle,

centrifuge at 10,000 rpm for 30 minutes at 4deg C in the RC5-B using the GSA rotor.

6. Add 6 ml of 5 mg/ml ethidium bromide, and cesium chloride such that the final concentration of cesium chloride is 1 g/ml.
7. Transfer the sample into 35 ml polyallomer centrifuge tubes and top off with a 1:1 solution of 100:10 TE buffer and cesium chloride, remove air bubbles, seal with rubber stoppers, and crimp properly.
8. Centrifuge at 60,000 rpm to 16-20 hours at 15-20degC in the Sorvall OTD-75B ultracentrifuge using the T-865 rotor.
9. Visualize the ethidium bromide stained DNA under long-wave UV light, and remove the lower DNA band using a 5 cc syringe and a 25 gauge needle. (It may be helpful to remove and discard the upper band first).
10. To remove the ethidium bromide, load the DNA sample onto an 1.5 ml Dowex AG (BioRad) column, equilibrated as before, and collect 0.5 ml fractions.
11. Pool fractions with an A260 of 1.00 or greater into 35 ml Corex glass tubes, add one volume of ddH2O, and ethanol precipitate by adding 2.5 volumes of cold 95% ethanol. Incubate at least 2 hours at-20degC, centrifuge at 10,000 rpm for 45 minutes in the RC5-B using the SS-34 rotor. Gently decant the supernatant, add 80% ethanol, centrifuge as before, decant, and dry the DNA pellet in a vacuum oven.
12. Resuspend the DNA in 10:0.1 TE buffer.

Single-stranded M13 DNA Isolation Using Phenol

This isolation procedure is the method of choice for preparation of M13-based templates to be used in Sequenase[TM] catalyzed dye-terminator reactions. A pre-incubated early log phase JM101 culture is prepared by transferring a thawed glycerol stock into 50 ml of liquid media and incubating for 1 hour at 37degC with no shaking.

M13 plaques are picked with a sterile toothpick and placed into 1.5 ml aliquots of the early log phase JM101 culture, which are incubated in a 37deg C shaker for 4-6 hours. After incubation, the bacterial cells are pelleted by centrifugation and the viral containing supernatant is transferred to a clean tube.

The phage particle are precipitated with PEG, collected by centrifugation, and the pellet is resuspended in buffer. The phage protein coat is denatured and removed by one phenol and two ether extractions. After ethanol precipitation, the dried DNA pellet is resuspended in buffer, and the concentration and purity crudely are assessed by agarose gel electrophoresis against known standards.

Protocol:

1. Prepare an early log phase culture of JM101, as above, and pick M13-based plaques with sterile toothpicks into 12 X 75 mm Falcon tubes containing 1.5 ml aliquots of the cells. Incubate for 4-6 hours at 37degC with shaking at 250 rpm.
2. Transfer the culture to 1.5 ml microcentrifuge tubes and centrifuge for 15 minutes at 12,000 rpm at 4degC.
3. Pipette the top 1 ml of supernatant to a fresh 1.5 ml microcentrifuge tube containing 0.2 ml 20% PEG/2.5 M NaCl to precipitate the phage particles. Mix by inverting several times and incubate for 15-30 minutes at room temperature.
4. Centrifuge for 15 minutes at 12,000 rpm at 4degC to collect the precipitated phage. Decant the supernatant and remove residual PEG supernatant by suctioning twice.
5. Resuspend the pellet in 100 ul of 10 mM Tris-HCl, pH 7.6 by vortexing, and add 50 ul of TE-saturated phenol.
6. Extract the DNA with phenol and twice with ether, as discussed above, and then ethanol precipitate.
7. Resuspend the dried DNA in 6 ul of 10:0.1 TE for use in single-stranded Sequenase[TM] catalyzed dye-terminator sequencing reactions.

Biomek-automated Modified-eperon Isolation Procedure for Single-stranded M13 DNA

This semi-automated method is a modification of a previously reported procedure, and allowed the simultaneous isolation of 48 single-stranded DNAs per Biomek 1000 robotic workstation within 3 hours. Basically, M13 plaques are picked with sterile toothpicks into aliquots of early log phase JM101, prepared as discussed above. The phage infected cultures are incubated in a 37degC shaker for 4-6 hours, transferred into microcentrifuge tubes, centrifuged to separate bacterial cells from the viral supernatant, and then carefully placed on the Biomek tablet. For each sample, two 250 ul aliquots are robotically distributed into two wells of a 96-well microtiter plate, and this process is repeated for each of the 48 samples until the entire 96 wells are filled. A solution of polyethylene glycol (PEG) then is added robotically to each well and mixed. The microtiter plate is covered with an acetate plate sealer, incubated at room temperature to precipitated the phage particles, and then centrifuged. The supernatant then is removed by inverting the plate and gently draining on a paper towel, without dislodging the pellet. After placing the microtiter plate back on the Biomek, a more dilute PEG solution is robotically added to each well. The plate then is covered with another sealer and centrifuged again. This rinse step aided in the removal of contaminating proteins and RNA. After removing the supernatant, as before, and placing the microtiter plate back on the Biomek, a Triton X-100 detergent solution is robotically added to each well. The plate is agitated gently and the sample from each pair of wells is robotically transferred to microcentrifuge tubes, which then are capped and placed in an 80deg C water bath for 10 minutes to aid in the detergent solubilization of phage coat proteins. After a brief centrifugation to collect condensation, the single-stranded DNA is ethanol precipitated, dried, and resuspended. An aliquot from each DNA sample is subjected to agarose gel electrophoresis to crudely assay concentration and purity. The yield of single-stranded template is approximately 2-3 ug per sample.

Protocol: The entire procedure will require 9 rows of P250 tips (counting from the centre of the Biomek tablet towards the left) for the isolation of 48 templates (48ISOL). The reagent module should contain PEG-2000, Triton-Tris-EDTA, and ethanol-acetate, respectively.

1. Prepare an early log phase JM101 culture in 50 ml of 2xTY, as above.
2. Using sterile toothpicks, transfer individual M13 plaques into 12 X 75 mm Falcon tubes containing 1 ml early log phase cell cultures, and incubate for 4-6 hours at 37degC with shaking at 250 rpm. (Growth for longer than 6 hours results in cell lysis and contamination of the phage DNA by cellular proteins and nucleic acids).
3. Separate the bacterial cells from the viral-containing supernatant by centrifugation at 12,000 rpm for 15 minutes at 4degC. Carefully open the tubes and place on the Biomek tablet.
4. The Biomek will distribute two 250 ul aliquots of viral supernatant per sample into the wells of a 96-well flat-bottomed microtiter plate (Dynatech). The Biomek then will add 50 ul of 20% PEG/2.5 M NaCl solution to each well, and mix by pipetting up and down.

5. Cover the plate with an acetate plate sealer and incubate at room temperature for 15 minutes.
6. Pellet the precipitated phage by centrifuging the plate at 2400 rpm for 20 minutes in a Beckman GPR tabletop centrifuge. Remove the plate sealer and drain the PEG from the plate by gently draining upside down on a Kimwipe.
7. Return the plate to the tablet, and the Biomek will robotically add 200 ul of PEG:TE rinse solution to each well. Cover the plate with a plate sealer, centrifuge, and drain, as above.
8. Return the plate to the tablet, and the Biomek will add 70 ul of TTE solution to each well. Remove and gently agitate to resuspend.
9. The Biomek then will robotically pool the contents from each pair of wells into 1.5 ml microcentrifuge tubes.
10. Incubate the tubes at 80degC for 10 minutes to denature the viral protein coat and then centrifuge briefly to reclaim condensation.
11. Ethanol precipitate the DNA by adding 500 ul ethanol/acetate to each tube, as described above.
12. Resuspend the DNA templates in 20 ul of 10:0.1 TE buffer.

96 Well Double-stranded Template Isolation

A manual as well as an automated procedure is given below. The automated method is a modification of a previously reported procedure which allows simultaneous isolation of 96 double stranded DNAs per Biomek 1000 Automated Laboratory Workstation within two hours. Basically colonies containing double-stranded plasmids are picked with sterile toothpicks into media and incubated at 37degC for 24 hours with shaking at 350 rpm. These cells are harvested by centrifugation and the pellets are either manually or robotically resuspended by the addition of TE-RNase solution. An alkaline lysis solution is used to lyse the cells and the lysate is precipitated with KOAc. The lysate is cleared by filtration and further concentrated by ethanol precipitation. An aliquot from each DNA sample is subjected to agarose gel eletrophoresis to crudely assay concentration and purity. The yield of double stranded template is approximately 3 mg per sample.

Protocol: Manual Double stranded isolation method: The following is a manual, 96 well, double stranded sequencing template isolation procedure that has been developed in our laboratory. A similar procedure that has been automated on the Biomek is presented elsewhere herein.

1. Pick individual shotgun clones off a plate with a steril tooth pick and deposit each separately into 96 well block containing 1.75 ml of TB media per well. Keep toothpick in media for about 5 minutes to allow the cells to defuse into the media, remove the toothpicks, cover the 96 well block with the loose fitting lid, and allow the cells to grow for 24 hours in the 37degC shaker/incubator at 350 rpm.
2. Remove block from the shaker/incubator and collect the cells by centrifugation at 2500 rpm for 7 minutes. The cells can be stored frozen at -20degC in the block at this stage.
3. After thawing the cells, add 100ul TE-RNAse-A solution containing RNAse T1, mix by pipetting up and down 4-5 times to resuspend the cell pellet and then incubate in the 37degC incubator/shaker for 5 minutes at 350 rpm to mix more thoroughly.
4. Remove the block from the incubator/shaker and then add 100ul of alkaline lysis solution. Shake the block by hand to mix the reagents and then incubate at room temperature for 1 hour with intermittent swirling. 5. Then add 100ul of either 3M potassium or sodium acetate, pH 5, and place the block in the 37degC shaker/incubator for 5 minutes at 350 rpm to thoroughly mix and shear genomic DNA to reduce the viscosity of the solution. Place the block at -20degC for 30 minutes.
6. Centrifuge the block in the GPR centrifuge at 3000 rpm at 4degC for 30 minutes.
7. Carefully remove 200 ul of the supernatant from each well in the 96 well block with the 12 channel pipetter and transfer them to a v-bottom microtiter plate, being careful not to transfer any cell debris.
8. Transfer 10 ul of supernatant into the respective cycle sequencing reaction tubes, and precipitate with 150 ul of 95% ethanol (without added acetate). After storage at -20degC for 30 minutes, the pellet is collected by centrifugation, washed three times with 70% ethanol, and dried directly in the cycle sequencing reaction tubes.
9. Prior to adding the fluorscent terminator cycle sequencing reaction mix, the dried templates should be stored at -20degC. An additional 75 ul of the supernatant is transferred to a Robbins PCR reaction tube (in 96 well tube format) and precipitated with 200 ul of 95% ethanol, washed three times with 70% ethanol, and stored dry at -20degC for future use.

The following is an automated, 96 well, double stranded sequencing template isolation procedure that has been developed in our laboratory.

1. Pick colonies using a toothpick into 1.8 ml TB with TB salts containing appropriate antibiotic and shake for 22-24 hours at 350 rpm in a 96 well block with cover.
2. Harvest cells by centrifugation at 1800 rpm for 7 min. Pour off supernatant and allow pellets to drain inverted. Cell pellets may be frozen at this point if necessary.
3. Turn on Biomek, begin the program DSISOL2 and set up the Biomek as indicated in the configuration function on the screen. Specifically, you should put TE-RNase solution in the first module, alkaline lysis solution in the second reagent module and 3 M KOAc, pH 4.8 in the third module.
4. Place the 96 well block containing cells onto the Biomek tablet at the position labelled "1.0 ml Minitubes". Place a Millipore filter plate in the position labelled "96well flat bottomed microtitre plate".
5. Press ENTER to continue with the program.
6. First the Biomek will add 100 ml TE-RNase solution to the cell pellets and mix to partially resuspend.
7. Next, the biomek will add 100 ml alkaline lysis solution to the wells of the filter plate.
8. The biomek then will mix the cell suspension again, transfer the entire volume to the filter plate containing alkaline lysis solution, and mix again. Set up the filtration apparatus with a clean 96 well block to collect the filtrate (wash and reuse the block used for growth).
9. The biomek will add 100 ml 3M KOAc, pH 4.8 to the wells of the filter plate and mix at the sides of the wells. Some choose to place the filter plate at -20degC for 5 minutes at this point. Transfer the filter plate to the QiaVac Vacuum Manifold 96 and filter using water vacuum only (do not do a harsh filtration as the plates are fragile and will loose their seal). This will typically take less than 20 minutes.
10. The supernatant collected in the 96 well block is the crude DNA and must be ethanol precipitated before use by the addition of 1 ml 100% ethanol and incubation at -20degC for at least 30 minutes.
11. Centrifuge for 25 minutes at 3000 rpm in a cooled Beckman GPR centrifuge.

12. Decant and wash with 500 ml 80% ethanol and centrifuge for an additional 5 minutes at 3000 rpm.
13. Decant the supernatant, drain inverted on a paper towel. Dry under vacuum.
14. Resuspend in 50 ml 10:0.1 TE for use in dye primer or dye terminator sequencing chemistry.

Genomic DNA Isolation from Blood

Genomic DNA isolation is performed according to the FBI protocol. After the blood samples (stored at -70degC in EDTA vacutainer tubes) are thawed, standard citrate buffer is added, mixed, and the tubes are centrifuged. The top portion of the supernatant is discarded and additional buffer is added, mixed, and again the tube is centrifuged. After the supernatant is discarded, the pellet is resuspended in a solution of SDS detergent and proteinase K, and the mixture is incubated at 55deg C for one hour. The sample then is phenol extracted once with a phenol/chloroform/isoamyl alcohol solution, and after centrifugation the aqueous layer is removed to a fresh microcentrifuge tube. The DNA is ethanol precipitated, resuspended in buffer, and then ethanol precipitated a second time. Once the pellet is dried, buffer is added and the DNA is resuspended by incubation at 55degC overnight, the genomic DNA solution is assayed by the polymerase chain reaction.

Protocol:

1. Blood samples typically were obtained as 1 ml of whole blood stored in EDTA vacutainer tubes frozen at -70deg C.
2. Thaw the frozen samples, and to each 1 ml sample, add 0.8 ml 1X SSC buffer, and mix. Centrifuge for 1 minute at 12,000 rpm in a microcentrifuge.
3. Remove 1 ml of the supernatant and discard into disinfectant.
4. Add 1 ml of 1X SSC buffer, vortex, centrifuge as above for 1 minute, and remove all of the supernatant.
5. Add 375 ul of 0.2M NaOAc to each pellet and vortex briefly. Then add 25 ul of 10% SDS and 5 ul of proteinase K (20 mg/ml H2O) (Sigma P-0390), vortex briefly and incubate for 1 hour at 55degC.
6. Add 120 ul phenol/chloroform/isoamyl alcohol and vortex for 30 seconds. Centrifuge the sample for 2 minutes at 12,000 rpm in a microcentrifuge tube.
7. Carefully remove the aqueous layer to a new 1.5 ml microcentrifuge tube, add 1 ml of cold 100% ethanol, mix, and incubate for 15 minutes at -20deg C.

8. Centrifuge for 2 minutes at 12,000 rpm in a microcentrifuge. Decant the supernatant and drain.
9. Add 180 ul 10:1 TE buffer, vortex, and incubate at 55degC for 10 minutes.
10. Add 20 ul 2 M sodium acetate and mix. Add 500 ul of cold 100% ethanol, mix, and centrifuge for 1 minute at 12,000 rpm in a microcentrifuge.
11. Decant the supernatant and rinse the pellet with 1 ml of 80% ethanol. Centrifuge for 1 minute at 12,000 rpm in a microcentrifuge.
12. Decant the supernatant, and dry the pellet in a Speedy-Vac for 10 minutes (or until dry).
13. Resuspend the pellet by adding 200 ul of 10:1 TE buffer. Incubate overnight at 55degC, vortexing periodically to dissolve the genomic DNA. Store the samples at -20degC.

Plasmids

A plasmid is an independent, circular, self-replicating DNA molecule that carries only a few genes. The number of plasmids in a cell generally remains constant from generation to generation. Plasmids are autonomous molecules and exist in cells as extrachromosomal genomes, although some plasmids can be inserted into a bacterial chromosome, where they become a permanent part of the bacterial genome. It is here that they provide great functionality in molecular science.

Plasmids are easy to manipulate and isolate using bacteria. They can be integrated into mammalian genomes, thereby conferring to mammalian cells whatever genetic functionality they carry. Thus, this gives you the ability to introduce genes into a given organism by using bacteria to amplify the hybrid genes that are created in vitro. This tiny but mighty plasmid molecule is the basis of recombinant DNA technology.

There are two categories of plasmids. Stringent plasmids replicate only when the chromosome replicates. This is good if you are working with a protein that is lethal to the cell. Relaxed plasmids replicate on their own. This gives you a higher ratio of plasmids to chromosome.

So how do we manipulate these plasmids?

1. Mutate them using restriction enzymes, ligation enzymes, and PCR. Mutagenesis is easily accomplished by using restriction enzymes to cut out portions of one genome and insert them into a plasmid. PCR can also be used to facilitate mutagenesis.

Plasmids are mapped out indicating the locations of their origins of replication and restriction enzyme sites.

2. Select them using genetic markers. Some bacteria are antibiotic resistant. While this is a serious health problem, it is a godsend to molecular scientists. The gene that confers antibiotic resistance can be added (ligated) to the gene you are inserting into the plasmid. So every plasmid that contains your target gene will not be killed by antibiotics. After you transfect your bacterial cells with your engineered plasmid (the one with the target gene and the antibiotic resistant marker), you incubate them in a nutrient broth that also contains antibiotic (usually ampecillin). Any cells that were not transfected (this means they do not have your target gene in them) are killed by the antibiotic. The ones that do have the gene also have the antibiotic resistant gene, and therefore survive the selection process.
3. Isolate them (such as with alkaline lysis)
4. Transform them into cells where they become vectors to transport foreign genes into a recipient organism.

There are some minimum requirements for plasmids that are useful for recombination techniques:

1. Origin of replication (ORI). They must be able to replicate themselves or they are of no practical use as a vector.
2. Selectable marker. They must have a marker so you can select for cells that have your plasmids.
3. Restriction enzyme sites in non-essential regions. You don't want to be cutting your plasmid in necessary regions such as the ORI.

In addition to these necessary requirements, there are some factors that make plasmids either more useful or easier to work with.

1. Small. If they are small, they are easier to isolate (you get more), handle (less shearing), and transform.
2. Multiple restriction enzyme sites. More sites give you greater flexibility in cloning, perhaps even allowing for directional cloning.
3. Multiple ORIs. It is important to note that two genes must have different ORIs if they are going to be inserted in the same plasmid.

R-Plasmids

R-Plasmids-It was long known that repeated exposure of bacteria to antibiotics results in the development of a resistant strain. Some bacterial cells undergo gene mutation which makes them resistant to the antibiotic. They live and multiply 10 the presence of the drug, giving rise to a resistant strain. This type of resistance develops gradually, and can be treated by administering antibiotics before a large number of resistant cells are formed.

Another type of resistance was observed with respect to the antibiotic gentamycin. Bacterial strains developed resistance to gentamycin only, a few months after its introduction. The gentamycin-resistant strain was to also found to be resistant to three other antibacterial agents, chloramphenicol, ampicillin and sulfonamide Growth of the strain with any susceptible strain of bacteria conferred antibiotic resistance to the latter within a few hours. Thus the antibiotic resistance was infectious.

The infectious antibiotic resistance factor or R factor consists of genes producing resistance to the four antibiotics and genes that facilitate the, transfer of the R factor. They are carried in plasmids called the R (for resistance) plasmids. When a strain carrying R plasmids conjugates with other bacterial cells, a copy of the R plasmid is transferred to the latter, giving it the property of drug resistance.

R plasmids were first discovered in a strain of Shigella in the late 1950s during an epidemic of dysentery in Japan. Shigella acquired drug resistance with the introduction of antibiotic therapy in 1946. By 1964, 50% of shigellas were resistant to all the four major drugs (sulfonamides, streptomycin, tetracycline and chloramphenicol) against dysentery. Plasmid-mediated drug resistance has serious medical impli-cations. Firstly contact with a drug-resistant non-pathogenic member of the normal gut flora can make a drug-sensitive pathogen resistant. Secondly, a pathogen may acquire new R plasmid resistances during epidemics. Plasmid DNA occurs in two forms called co-integrates and aggregates. The plasmid co-integrate contains all the resistance genes, together with the genes required for transfer, in one circular molecule, e. g. RIGO (Cm, Sm Tu, Tc) isolated from Shigella flexneri.

In plasmid aggregates the various genes are located, at least for some time, on separate molecules, e. g. in one isolate of Salmonella typhimurium type 29.

Chapter

5

Disease Diagnosis and Control

Someone who has a plant problem asks two questions: What is the problem, and how do I correct it? These are hard questions for beginners and experts alike. Diagnosis is the process of determining the cause of a problem. It can be a long or short process depending on one's ability and the nature of the problem. Once the cause is known, an appropriate control strategy can be developed.

Disease Diagnosis

Plant pathologists take many different approaches to diagnosing plant disease problems. The first step is to decide whether the problem *is* a plant disease. The broadest definition of plant disease includes anything that adversely affects plant health. This definition can include such factors as nutrient deficiencies, lawnmower damage, air pollution, and pathogens. A stricter definition usually includes a persistent irritation resulting in plant damage. This excludes mechanical damage such as lawnmower injury to trees or natural events such as hail or lightning. A very strict definition includes only those (living) things that replicate themselves and spread to adjacent plants. This includes such biological organisms as nematodes, fungi, bacteria, and viruses. Plants damaged by macroscopic organisms, such as deer, rodents, and birds usually are not considered to be diseased.

Many novices use the picture-book method of diagnosis: looking at textbook pictures of problems and attempting to match the problem with the picture. "The Ortho Problem Solver" and the APS Plant

Disease Compendia series are examples of texts that have many useful, high-quality colour pictures. Although this method is useful for simple and common problems, it is usually inefficient and inaccurate for more complex or difficult problems. Another simple technique, used by "The Ortho Problem Solver," is the checklist. Through a series of 70 questions, a person builds a case history of the problem. The questions include the kind, age, and size of the plant. The plant's location, location of the property, and relationship to other plants also are part of the checklist. Information on the recent weather and soil conditions, soil coverings, and recent care also are needed. Describing the overall condition of the plant is very important.

Symptoms and signs are used to diagnose the condition of a plant. *Symptoms* are the physical characteristics of disease expressed by the plant. Symptoms can include wilt, galls, cankers, rots, necrosis, chlorosis, and general decline. Definitions of these can be found in the glossary. *Signs* are physical evidence of the pathogen causing the disease. Signs can include fungal fruiting bodies (such as mushrooms or pycnidia), mycelia, bacterial slime, presence of nematodes or insects, or the presence of insect holes accompanied by sawdust or frass. Again, these terms are defined in the glossary.

The checklist approach is a good guide for the types of questions to ask, but not all the questions are necessary. Some pertain only to certain situations or plants. Other approaches to disease diagnosis attempt to narrow the possibilities with each question and are like synoptic or dichotomous keys. The diagnostic keys differ depending on the author. Some keys start by identifying the part of the plant that is affected. Is it the whole plant, or just the leaves, branches, stems, or roots? Some refine the first questions to stages of the plant; for example, are the seedlings, flowers, fruit, or seed affected? Some diseases affect only certain portions of a plant. Many other references start out with the plant itself: is it a cherry tree, foliage plant, or zinnia? Only a limited number of diseases attack a given plant species.

There is no one key set of questions or techniques for diagnosing plant diseases. Experience and practice are the best teachers. It is easier to diagnose plant problems by making a personal, on-site inspection. Subtle influences of the site, plant environment, and possible management practices can be seen that may have been overlooked by the grower. Difficulties arise when the diagnostician is presented only a portion of the plant because that portion may or may not indicate the real problem.

The worst situation is a request for a diagnosis by phone, because misunderstandings and an inaccurate diagnosis can easily occur. However, sometimes this is the only contact someone may have with a diagnostician.

The Systematic Approach

To organize your approach to a plant disease problem, the "Systematic approach to diagnosing plant damage" by J. L. Green, O. Maloy, and J. Capizzi is useful and is presented briefly here. It consists of a detailed handout, a study guide, and a colour slide set. It is designed as a master gardener seminar and is useful to hear or to present yourself. The approach involves defining the real problem and distinguishing between living and nonliving causes of plant damage by looking for patterns, determining the development of the damage, and building a case history of the problem. With these steps, it is usually easy to narrow the possibilities and to turn to appropriate reference materials including textbooks, herbarium samples, and knowledgeable specialists.

Define the Real Problem Identify the plant and what it should look like at this time of year. A grower or gardener may mistake a normal stage of development for a diseased plant. Describe the abnormality in terms of *symptoms* and *signs*. Although a plant may exhibit symptoms of wilting, the real problem may be due to rotted roots, a girdled trunk, or lack of water. Determine what part(s) of the plant is/are affected. The rest of the procedure involves distinguishing between living and nonliving factors. Look for Uniform or Nonuniform Patterns Uniform damage is indicative of nonliving factors. Damage may occur on many plant species in the same area, on all the plants in a particular row or block, on all the leaves or shoots on one side of the plant, or on the same-age portion of each leaf or needle. Nonuniform damage to plants is indicative of living factors such as pathogens or insects. This damage shows up as scattered affected plants among a community of plants, scattered leaves or shoots on a single plant, or scattered spots on a single leaf.

Determine the Time Development of the Damage If damage does not spread or there is a clear line of demarcation between damaged and nondamaged tissues, this is indicative of nonliving factors. Spread of the damage from plant to plant or to other plant parts over time indicates damage by a living organism. Look for Specific Symptoms and Signs Look for signs such as fungal fruiting bodies, mycelial threads, bacterial slime, presence of insects, mites, or holes with frass.

Look for (or ask questions about) symptoms of nonliving factors that may be caused by extremes of temperature, light, or water. Ask about mechanical factors or chemical factors as indicated by uniform patterns. Check references for probable diseases of the identified plant. You may have to send the sample to an appropriate laboratory to continue to identify possibilities. Once you have determined that a real problem exists and is caused by a living organism, you need to decide what type of organism may be causing the damage. There are many fungal and viral diseases of plants and a few caused by bacteria and nematodes. Some insect problems can mimic diseases; however, this next discussion will focus only on pathogens.

An article entitled "Preliminary Diagnosis of Plant Diseases by Visual Symptoms" by C. Leach is useful and is briefly described here. This discussion focuses on biological causes; however, many physical factors such as extremes in environment also are discussed in the article. Begin by establishing which plant part or growth stage is showing symptoms. Are symptoms showing on roots, tubers, bulbs, corms, seedling, foliage, stem, branches, trunks, flowers, fruit, or on the entire plant? Often, you must next decide whether the symptoms are on the outside of the plant or whether you need to cut into it to see the symptoms.

Root Symptoms

External root symptoms include galls, discoloration, or death of roots or parts of roots. Crown gall caused by the bacterium *Agrobacterium tumefaciens* is common on many plants. Some fungal diseases, such as club root of cabbage, also cause galls. Root-knot nematodes, *Meloidogyne* spp., can cause large or small irregular galls.

Small discolored, dead areas may be caused by a large variety of fungi and root-lesion nematodes, *Pratylenchus* spp. General death of the entire root system or just feeder roots is indicative of many fungi. Remember, injury to the root system often includes yellowing, stunting, or wilting of aboveground parts. Many fungi, such as *Verticillium* and *Fusarium*, will cause an internal vascular discoloration as will some bacterial wilts.

Symptoms on Storage Organs (Tubers, Bulbs, Corms, etc.)

Discolored or dead areas that penetrate deep into the storage organs are caused by many fungi and some bacteria. Dry rots are often caused by fungi which may produce mycelia or spores. Soft rots are usually associated with bacteria such as *Erwinia* and can be accompanied by strong, repulsive smells. Many times bacterial soft rots will closely follow rots caused by fungi, making diagnosis difficult. Scurfy dead tissue on the surface may be caused by a variety of myxomycetes, such as powdery scab of potato. The filamentous bacterium *Actinomyces scabies* causes common scab of potato. Galling of storage organs can be caused by both fungi and nematodes. Internal problems, such as ring rot of potato, can be caused by bacteria fungi or by several viruses.

Seedling Diseases

If seedlings fail to emerge, or fall over and die, this is usually referred to as damping-off. Fungi such as *Rhizoctonia, Pythium,* and *Fusarium* are common and affect seedlings just at or below the soil line. Dead areas on cotyledons, the first leaves or stems, are usually caused by fungi but occasionally by bacteria. Spots caused by fungi may contain hyphae or fruiting bodies while spots caused by bacteria may appear water soaked around the margins. General yellowing, or mottled or patterned yellowing, can be caused by several viruses. Ringspots, blotches, and streaks of yellow or green may also occur with viral infections. Nematodes such as *Ditylenchus* can cause swelling or galling of stems or leaves. Rusty red, brown, or black spots or stripes are indicative of rust and smut fungi. White moldy growth can be caused by powdery mildew or downy mildew fungi.

Leaf Symptoms

Discoloration (yellowing or shades of green), which is localized or in distinct patterns, usually indicates a virus. Other leaf symptoms may occur with viral infections. A general or uniform yellowing may indicate a root rot of some kind, so remember to examine the entire plant. Dead areas on leaves can be caused by fungi or bacteria. Necrotic areas caused by fungi may contain hyphae or fruiting bodies, particularly after incubation in a warm, moist environment. Necrotic areas caused by bacteria may be distinguished by water-soaked margins or bacterial streaming. Small rusty red, brown, or black spots or stripes may be caused by rust and smut fungi. Distinctive spores can usually be found in these spots. Leaf distortion (elongated, dwarfed, etc.) can be caused by several viruses. Leaf galls usually are caused by fungi, such as peach leaf curl and azalea leaf gall. Viruses and nematodes rarely cause galls on the leaf. Moldy white appearance of leaves indicates powdery or downy mildew. Wilting indicates lack of water which may be due to one of the vascular wilt fungi, root rots, or bacterial wilts. Other parts may need to be examined for an accurate diagnosis.

Stem, Branch, and Trunk Disorders

The leaf disease remarks above pertain to most stem disorders of annuals or herbaceous perennials. Complete or partial death of woody stems or branches, usually referred to as cankers, can be caused by a large variety of fungi and a few bacteria. Cutting into the wood with a knife may reveal a sharp border between healthy and infected tissue. Look for spore-bearing structures of fungi or induce sporulation in a moist chamber. Some bacterial cankers will excrete a sticky ooze in the spring. Wood rots and decays are caused mainly by fungi which may produce large conks or bracketlike fruiting structures. Galling can be produced by the crown gall bacterium *Agrobacterium tumefaciens*. A few galls or swellings are caused by fungi, such as white pine blister rust and *Exobasidium* sp. Witches' brooms (excessive branching) can be caused by several fungi and by a group of parasitic plants called mistletoes.

Flower Symptoms

Abnormal colour changes and/or distortions can be caused by several different viruses. Partial or complete death of flower parts can be caused by fungi and bacteria. Fungi usually produce characteristic spores; bacterial infections can appear water soaked. Smut of grasses convert individual flowers or whole inflorescences into masses of black spores. Ergot, stamen blight of blackberry, and anther mold of red clover are other fungal examples of converted flower parts. Some nematodes can also convert flower parts for their own use.

Fruit Symptoms

Fungi cause a wide variety of decays, rots, and superficial spotting or russetting. Important symptoms include specific colour of rotted tissue, firmness of the tissue, and signs such as spores or spore-bearing structures. Viruses can cause discolorations and malformations. Bacteria may cause discrete spots on fruit in certain field situations or soft rots in storage.

Principles of Plant Disease Control

After you diagnose a plant disease, you are only half finished. The equally challenging task of designing the proper control recommendation is next. Understanding the specific disease or the life cycle of the pathogen involved is necessary to make an adequate control recommendation.

The Disease Triangle

Three major factors contribute to the development of a plant disease: a susceptible host, a virulent pathogen, and a favorable environment. A plant disease results when these three factors occur *simultaneously*. If one or more of these factors do not occur, then the disease does not occur. The genetic makeup of the host plant determines its susceptibility to disease. This susceptibility or resistance may be determined by various physical and biochemical factors. Plant stature, growth habit, cuticule thickness, and stomatal shape are a few physical factors that influence disease development. The plant's developmental stage also may influence disease development. Pathogens differ in their ability to survive, spread, and reproduce. Environmental extremes of temperature, light, or moisture can accentuate many diseases. Cool, moist conditions are ideal for many fungal pathogens.

The Disease Cycle

Understanding the disease cycle is important when considering control options. Learning the chain of events that contribute to a disease helps point out the weakest links. Control measures can then be used to break the cycle. Most pathogens must survive an adverse period, usually winter, when they do not actively incite plant diseases. This overwintering inoculum reinfects or continues infecting the plant host in the spring. Some diseases are characterized by a single cycle during the year. Other diseases continually produce new inoculum, repeating the cycle many times during the course of a single growing season.

Disease Control

The five basic principles of plant disease control are: exclusion, avoidance, eradication, protection, and resistance. These principles work at federal, state, county, and personal levels.

Exclusion This includes quarantines, inspections, and certification. Plant material is examined to prevent entry of a disease that does not already occur in a particular country, state, or geographic area. The most notable use of this method is in California. As you cross the state border, you must stop at an agricultural inspection station because they screen for fresh fruit, vegetables, and plants coming from areas known to have certain diseases (or insects). There are many quarantines on importing plant materials into the country as well as into the Pacific Northwest. For example, several grape viruses, Dutch elm disease, chestnut blight, eastern filbert blight, azalea flower spot, rust-susceptible varieties of barberry, mahonia (Oregon grape) or mahoberberis, and several peach viruses all are quarantined to prevent import into Oregon.

Quarantines are effective but may not always protect an industry. Quarantines on the movement of *Corylus* sp. infected with eastern filbert blight into the Pacific Northwest from the east coast were effective for more than a century. In 1972, an infection centre was discovered in Washington and has since moved into major production areas of Oregon. Avoidance If the disease does occur in your area, there may be techniques to avoid disease development. Choice of planting site, time of planting, storage conditions, or avoiding wounds are a few of these techniques. Phytophthora root rots can be avoided by not planting in heavy, poorly drained soils. Planting later in the year when soils are dryer and warmer will avoid some damping-off diseases common to many vegetables. Wounding can cause entry points for pathogens or weaken a plant to the point that it cannot defend itself. Avoiding wounds also helps to control the crown gall bacterium which needs an injury to begin the infection process. Planting certified virus-free stock is a good way to avoid virus-related diseases.

Good horticultural practices such as proper fertility, pruning, watering, and proper training will go a long way to help control plant diseases.

Eradication When a plant is infected or an area is infested with a plant pathogen, eradication can eliminate or reduce the disease threat. Rotation, sanitation, heat treatment, eliminating the alternate host, and certain chemicals can be used to reduce or eliminate diseases. Crop rotation is a common method in both commercial agriculture and home gardens. It is necessary to know the pathogen and its host range. Rotation reduces soil populations of fungi or nematodes only if nonhost plants are used.

Removing plant debris (sanitation) is important where pathogens may overwinter. Raking leaves, removing rotted fruit, picking up old vines, and pruning out dead wood or canes all are part of sanitation.

Once collected, dispose of debris by burning, burying, or hot composting. If you decide to compost, it must be done correctly or completely or you will not benefit from your efforts. Field burning is another method of sanitation which destroys grass stubble where plant pathogens may overwinter.

Rusts are a group of fungi that can complete their life cycle on two or more different hosts. Eliminating an alternate host may help reduce pressure from these diseases. Heat treatment is usually used to eliminate viruses from propagation material.

Certain chemicals can be used to eliminate infections or infestations. Soil can be fumigated to reduce populations of certain fungi and nematodes. Lime sulfur is used during the dormant season to denature and kill fungal fruiting bodies or spores. Some fungicides have kickback activity, meaning that infections of some fungi can be stopped if the chemical is applied within a few days after the infection has started.

Protection Protection is treating a healthy plant before it becomes diseased. There are both biological and chemical means of protection. One of the most successful biological protections has been the use of a bacterium to protect against a bacterial disease known as crown gall. The roots of a seedling or nursery plant are merely dipped into a suspension of a commercial preparation of the bacterium prior to planting.

Chemical protection is one of the most widely used means of control. Some fungicides (such as copper and sulfur products) are allowed for use under several "organic" growing guidelines. Many fungicides are on the market but few can be obtained easily by the homeowner. It is necessary to know the disease cycle and host susceptibility to get good control using fungicides. Proper timing, coverage, and selection of fungicides is also needed.

Resistance Resistance is a term sometimes mistakenly used interchangeably or in conjunction with "immunity," "tolerance," and "susceptibility." These terms describe the inherent genetic makeup of the plant and thus its reaction to plant pathogens. Resistance and its opposite, susceptibility, are levels or degrees of a plant's reaction. Some cultivars of a plant can be more or less resistant (or susceptible) than another cultivar. Resistant cultivars can still become diseased but not as much as (or more than) another. If a plant does not ever become diseased, then the term "immune" can be used. Tolerance describes a plant (usually a food crop) that may become diseased but produces yields similar to a healthy plant.

Lists of resistant plants can be found in many texts and seed catalogues. Planning ahead is essential and planting resistant cultivars is the easiest means of disease control.

Knowing what diseases a plant is susceptible or resistant to can help in the diagnostic process. One can eliminate possibilities by knowing which diseases are likely to occur.

Summary

Experience and practice are the best teachers of plant disease diagnosis. Examination of the plant's physical environment and management history are essential. Observing patterns and specific symptoms and signs are important in arriving at a correct diagnosis. Once diagnosed, the proper control measures can be formulated. Knowledge of the host, pathogen life cycle and environmental factors also aid selection of the most effective control measures. A combination of exclusion, avoidance, resistance, eradication, and protection will control most plant diseases.

Genomics of Plant-Associated Bacteria: A Glimpse of the Future that has Become Reality

The largest amount of biochemical and genetic diversity on Earth exists within the prokaryotic domains Archaea and Bacteria. Members of these domains are found in every conceivable (and sometimes inconceivable) environmental niche based on their ability to utilize diverse energy sources, electron donors and acceptors (8). Because higher organisms on Earth evolved in the presence of prokaryotes, they are intimately associated with every organism. Prokaryotes are responsible also for most of the major steps of biogeochemical cycles without which plant, animal and human life would be impossible.

A subclass of prokaryotes of direct interest to plant pathologists includes those bacteria that interact intimately with plant hosts and directly affect overall plant health and survival. Some prokaryotes serve to improve plant health while others negatively impact plant health, ranging from slightly deleterious effects to partial or complete death of the plant. The development of the technology to sequence large stretches of DNA (genomics) and to analyze this DNA (bioinformatics) is beginning to have enormous impacts not only on the way scientists carry out research but also in the way we think about the overall biology of plant-microbe interactions.

History

It can be argued that the field of genomics actually began several years ago in virology, where the ability to determine the DNA (or RNA) sequence of a virus allowed comparisons of genomes, enzyme activities, and many other aspects previously difficult or impossible to perform. The net effect of the development of viral genomics was the revolution of plant virology and the resultant explosion in our understanding of plant-viral interactions.

In an analogous manner, the emergence of microbial genomics will enable us to answer classical questions about plant-associated microbes that once seemed, at least to some of us, as unobtainable as intergalactic space travel. In the pre-genomics era, we probed the microbial universe one factor at a time, adding incrementally to our knowledge of one or a few bacteria. As we enter the age of high throughput sequencing, or what is referred to as "the age of genomics," scientists can begin to use a holistic approach for answering questions of consequence to plant health. Some of the many exciting questions are:

- What is the minimal set of genes required for survival on or in a plant?
- What is the arrangement of pathogenicity and virulence genes in plant pathogenic bacteria?
- Are genetic determinants of host range or host preference clustered in the genome?
- Are there distinctive features that will predict a bacterium's ability to survive on or in different plant parts?

As recently as 1995 The Institute for Genomic Research (TIGR) published the first complete sequence of a bacterial genome. Since then the number of completed bacterial genomes has risen to 24. Several others are expected to be completed within the next 2 years. Most of the genome sequencing efforts so far have targeted medically or industrially important microbes and relatively few target plant-associated bacteria. Fortunately, genome sequencing efforts in Brazil, France and the United States increasingly focus on plant pathogenic bacteria. In their very recent article, a team of Brazilian scientists from São Paulo present the first complete genome sequence of a phytopathogen. Simpson et al., sequenced the entire genome of a strain of *Xylella fastidiosa*, and will soon finish sequencing another plant pathogen, *Xanthomonas axonopodis* pv. *citri*. In 2000, the United States Department of Agriculture (USDA) launched the first US public initiative for complete genome sequencing of agriculturally related microbes, including plant-associated bacteria (IFAFS). Together with private ventures in the US, France and elsewhere, the amount of genomic information on plant-associated bacteria is increasing exponentially.

APS Symposium on "Genomics of Plant-Associated Bacteria"

The age of microbial genomics and its implications for plant health will be highlighted at the upcoming 2000 Annual Meeting of the

American Phytopathological Society (August 12-16, 2000, New Orleans) in a symposium entitled "Genomics of Plant-Associated Bacteria". The APS Bacteriology Committee is sponsoring the symposium. Co-organizers for the symposium are Leland S. Pierson III, University of Arizona, and Carol A. Ishimaru, Colorado State University. An international group of speakers will present the current status and future goals of several essential microbial genome sequencing projects. The symposium will be held Tuesday, August 15, 2000 from 1:00 to 5:00 PM. The symposium is supported by funds provided by the APS Council, the College of Agriculture and Life Science at the University of Arizona, and the Colorado Agricultural Research Station and College of Agricultural Sciences at Colorado State University. Funds for the Symposium are provided also through a work agreement between the Animal and Plant Health Inspection Service (APHIS)/USDA and the APS Bacteriology Committee, which resulted in a summary of plant pathogenic bacteria prevalent in the U.S.

Generation of Sequence Data

The genomes of plant-associated bacteria contain DNA, which is made up of the four nucleic acid bases, adenine (A), guanine (G), thymine (T) and cytosine (C). It is the specific number and order of these bases that uniquely identifies an organism. Bacterial genomes consist of large DNA molecules arranged in one or more circular or linear chromosomes.

In addition to chromosomes, bacterial genomes often contain extra chromosomal DNA molecules, called plasmids, which may vary in length from a few kilobases (kb) to over 400 kb. Estimated genome sizes of plant pathogenic bacteria vary among species. Some of the smaller genomes are estimated to be in the range of 2.5 million (mega) base pairs (Mbp) and the larger ones are probably greater than 5 Mbp.

The goal of genome sequencing is the production of a map that identifies the linear arrangement of all bases in an organism's DNA. Obtaining such a map is not a trivial undertaking; however, technical advances in sequencing and data processing continue to reduce the time and expenses involved. The complete sequencing of a 2.5-Mbp genome presently can take less than six months at a cost of less than $500,000.

Obtaining the complete sequence of an organism usually involves multiple sequencing strategies. A common approach is to use a combination of shotgun-sequencing, in which small fragments of the DNA are cloned and sequenced, followed by a gap-closure phase. In the latter, larger DNA fragments contained in cosmid or BAC libraries are sequenced. When finished, the complete sequence gives the exact size and structure of the genome, including the number and form of chromosomes and plasmids.

Sequence Annotation (Bioinformatics)

To make biological sense out of all the As, Ts, Gs, and Cs, sequence data must be "mined" or analyzed by software specifically developed to identify possible open reading frames (ORFs). While ORFs are associated with sequences that are transcribed and translated into proteins, other sequences give information about gene expression. The term bioinformatics describes the mining of biological sequence information to predict gene function, protein and RNA structure, gene regulation, genome organization and phylogenetic history of genes and gene families. During this process, a test sequence is compared to all the sequence information deposited in worldwide gene banks. Database searches provide a probability score that one sequence is similar to another. If highly significant, these similarities may infer homology (relatedness through evolution) or function. The result of all this data mining is an annotated map of the genome.

Comparative Genomics: An Emerging Field

Recent software developments enable analysis of the extremely large data sets generated by complete genome sequencing and provide for comparisons at the whole genome level. As an example, the program MUMmer (Maximal Unique Match) aligns segments of closely related DNA sequence several million base pairs in length. This level of data mining or "comparative genomics" has direct and exciting implications for plant pathology. As the number of complete genome sequences increases, comparisons across genera, species, subspecies, pathovars, and strains will become possible.

What can we Learn from a Genomics Approach?

Through genomics, we can begin to address questions about the evolution of plant-associated bacteria. By comparing the arrangement of DNA along entire genomes, we can begin to propose models of how genetic information passed between organisms. Envision the magnitude of this new perspective as the difference between looking at the earth from your back porch and looking at it from a space shuttle. All inversions, rearrangements, and insertions in an organism's genome would be viewable, giving rise to all sorts of questions: What does the arrangement of genes in a pathogen tell us about evolution of host specificity or pathogenicity? Which sequences indicate gene transfer between related or unrelated organisms? The range of exciting research questions being limited only by the imagination.

The complete genome sequence gives the exact proportion and distribution of each of the four bases in the genome. The relative amount of each base, reported as the G+C content, is characteristic of a genus. Based on prior measurements, the genera of certain plant pathogenic bacteria, such as *Erwinia*, are predicted to have moderate G+C contents, while others, such as *Clavibacter* and *Pseudomonas*, are predicted to have much larger G+C contents (upwards of 70%).

Localized variations in the G+C content along the genome have significant implications for understanding the potential routes of genetic exchange between organisms. For example, by using genetic tools such as transposon mutagenesis and gene complementation, plant bacteriologists have located and sequenced large segments of DNA associated with pathogenicity. The G+C contents of some of these pathogenicity-related regions are distinct from the overall G+C content of the organism.

From this and other information, scientists postulate that the presence of "pathogenicity islands" can indicate how some bacteria may have evolved to become pathogens.

With a major emphasis on large-scale sequencing of plant-associated bacteria will come knowledge of the functional relationships between bacteria, especially those with different pathogenic life styles. Perhaps we will find evidence that genomes of bacteria from similar habitats contain similar sets of genes required for survival in that habitat. Perhaps we'll identify regions that are distinctive of life as an epiphytic or life as an endophyte.

Surely, the era of genomics will usher in significant advances in our understanding of plant-microbe interactions. A few of the many possible breakthroughs will be in:

- Identification of new surface proteins that play critical roles in plant-microbe interactions;
- New insights into unique biochemical pathways critical for plant-microbe interactions;
- Identification of requirements for culturing fastidious microorganisms such as phytoplasmas;
- New targets for specific and sensitive detection of plant-associated bacteria.

Changing Science and Plant Pathology Forever

The World Wide Web affords immediate access to all the available information contained in an organism's DNA. This access results in near real-time feedback and analysis by the entire scientific community. As Dr. Stanley Falkow eloquently said recently, "[Genomics] reminds me of the 'good old days' of bacterial conjugation when each day's experiments revealed some new facet of genes and biology.... There was just too much to discover to worry about competition". Genomics challenges our academic institutions to provide new graduate education experiences that can advance the use of these strategies and develop new partnerships between disciplines. Already disciplinary lines are becoming blurred and discipline insularity is becoming less prevalent.

What's Next?

Once the complete genome sequence of an organism is known and annotated, the next challenge is identifying gene functions of all possible ORFs. Plant pathologists will provide the human intelligence needed to analyze and rank the importance of the ORFs for their possible biological significance.

Bioinformatics approaches will identify functions of a portion, but not all, of the ORFs in a genome, and some of the assigned functions may actually be incorrect. In *Escherichia coli*, the common intestinal bacterium now studied intensively for over forty years, the functions of one-third of its 4,288 ORFs (*ca.* 1,415) are still unknown.

Thus, gene disruption and replacement experiments will continue to play a critical role in functional analyses of predicted ORFs. Still, the shear volume of sequence data generated by genome projects demands improvements be made in technologies for functional analysis. One such technology is DNA microarrays, which have quickly become the method of choice for examining gene expression on a genomic scale.

The Ultimate Goal

With the development of microbial genomics comes a responsibility to plant pathologists of making the most of these powerful resources. We must begin applying this information for reducing threats and losses from plant pathogenic bacteria and for improving the viable use of beneficial bacteria. There are very few means of controlling plant pathogenic bacteria. The limited number of chemicals and sources of disease resistance means the best strategy is still prevention through exclusion. Plant virologists have already utilized their knowledge of complete plant viral genomes to develop transgenic virus-resistant plants. Will we be able to do the same with bacterial genomics? Perhaps other methods of disease control will develop in a manner analogous to pharmaceutical approaches for veterinary and clinical medicine. Knowing the complete genome sequences of many plant-associated bacteria may allow us to identify the Achilles' heel of a plant pathogenic bacterium or alternatively to convert a potential silver bullet to a golden one.

Bacterial Wetwood and Slime Flux of Landscape Trees

Bacterial wetwood, a water-soaked condition of wood, occurs in the trunk, branches, and roots of many shade and ornamental trees, but is often not obvious in trees less than 10 years old. Wetwood is most prevalent and causes the most damage to elms, especially older elms (including American [*Ulmus americana*], Chinese [*U. parvifolia*], English [*U. procera*], European [*U. carpinifolia*] Siberian [*U. pumila*] slippery [*U. rubra*], and winged [*U. alata*] and poplars). Wetwood wilt and dieback of elms are most common in the Midwest and Great Plains. It is associated with the death of branches in large cottonwoods and with the dieback and premature death of Lombardy poplars, some as young as two years. Most trees die back slowly over a period of several years or more, however, small trees may wilt and die suddenly.

Symptoms vary with location suggesting that environment influences disease development. Trees in the western United States show more variability in internal disease development and much less bleeding and symptom expression than trees in the East and Midwest. Wetwood is normally not serious in most trees but as a chronic disease it can contribute to a general decline in tree vigour, especially of older trees growing under adverse conditions. Wetwood in landscape trees is usually unimportant except for the disfiguring appearance of light or dark streaks where liquid seeps out of cracks and wounds and flows down the bark. As the bleeding occurs, the liquid or flux flows down the trunk, wetting and soaking large areas of bark.

As it dries, a light gray to white incrustation is left. If the exudation continues over a period of time, the bark has a two-coloured vertical column of brown with strips of white at the margins. This liquid is toxic and commonly causes localized death of the cambium at the base of a pruning cut and around trunk cracks. The liquid on the bark surface becomes contaminated with mixed populations of many different kinds of airborne bacteria, yeasts, and filamentous fungi that give it a slimy texture and often a fetid odor. The gray to brown, foamlike foul liquid is called slime flux or wetwood slime. The slime flux may prevent the healing of wounds by retarding or preventing callus formation. Fluxing occurs in Illinois from April to December but is most conspicuous during the summer, ceasing during the winter.

Fluxing sap of wetwood-affected tree-sufficiently toxic to prevent callus formation and kill bark at base of pruning wound (IL Natural History Survey photo). Slime flux is distinct from the white or "alcoholic" flux that seeps from sites where various bacteria and other organisms infect shallow, localized wounds in the inner bark and cambial region made by insects, ice, wind, lightning, or lawn mowers, pruning tools, and machinery. The frothy alcoholic flux is acidic, nearly colorless, often gives off a pleasant fermentative odor, and only persists for a short time in summer. Alcoholic flux is not related to wetwood.

Causal Organisms

In some trees, especially elms and poplars, bacteria are consistently associated with wetwood and apparently cause it. In elms, the following bacteria are commonly isolated from diseased wood: Enterobacter cloacae (formerly *Erwinia nimipressuralis*), *Enterobacter agglo-merans* (synonym *Erwinia herbicola*), *Bacillus megaterium, Pseudomonas fluorescens*, and *Klebsiella oxytoca*. Elm bark associated with bleeding wounds harbours large populations of bacteria that may be disseminated by various means to infect other stem or branch wounds; Enterobacter cloacae has been isolated from the native elm bark beetle (*Hylurgopinus rufipes*), a vector of the Dutch elm disease fungus. Infected pruning tools may also transmit bacteria from wetwood-infected to healthy tissue.

Symptoms

Within the trunk and larger limbs of elms, poplars, and other trees, wetwood appears as a dark brown to black water-soaked area which in cross section generally appears more or less circular. Wetwood liquid under pressure sometimes spreads to the outer sapwood. In elms and poplars it may cause gray-brown streaks or broken bands in one or more annual rings. Most wetwood in the large branches and trunks of trees appears confined to a discolored central zone and to the innermost sapwood free of discoloration, but may extend almost to the cambium in wounded stems. In young elms, from 15 to 30 years old, the central core of wetwood-affected tissue tapers to a point just above the stem-root interface. It is also as a dark brown discoloration in root tissues, often associated with a wound, and extending upward toward the main stem. It does not usually occur in grafted roots.

Bacterial wetwood in broadleaved trees is differentiated from normal sapwood or heartwood by its visibly darkened colour, a sour or rancid odor due to an accumulation of fatty acids, a higher moisture content, increased alkalinity (by one pH unit or more), decreased electrical resistance, abnormally high gas pressure, a higher mineral content and specific gravity, less oxygen and more carbon dioxide. It often contains methane gas. Liquid from elm wetwood may have up to 11 times more calcium, magnesium, and potassium cations than healthy sapwood. Extreme gas pressures, up to 60 pounds per square inch (4.22 kg/cm^2) have been detected in discolored wetwood in elms, while gas pressures up to 0.7 kg/cm^2 are common.

Pressures fluctuate seasonably with the highest readings from May through September. The high gas pressures frequently force liquid to seep from wounds and flow down the bark. Gas is produced in wetwood-affected trees from the metabolic activity of bacteria. The gas typically consists of 45 to 60 percent methane, 0 to 7 percent oxygen, 6 to 16 percent carbon dioxide, 23 to 34 percent nitrogen, and 1 percent hydrogen.

The air in normal wood contains about 19 percent oxygen, 75 percent nitrogen, and up to 5 percent carbon dioxide. Methane and hydrogen are produced by certain bacteria under anaerobic conditions and do not occur in normal wood. *Enterobacter cloacae*, and probably other bacteria, ferment carbohydrates, producing carbon dioxide which is converted to methane by further bacterial action. The nearly anaerobic condition of wetwood prevents decay by wood-rotting fungi.

The first external sign of wetwood is usually the bubbling and seepage ("bleeding" or fluxing) from wounded tissue in V-shaped branch crotches, wounds made by removal of branches, injection holes (for fungicides, insecticides, or nutrient uptake), and trunk cracks, ribs, or beaks. Internal gas pressure commonly reopens old wounds. The sour liquid is colorless to tan as it oozes out and may contain up to 10 billion bacteria per milliliter.

Foliage in the tops of trees severely affected by bacterial wetwood sometimes curl, scorch, droop, turn yellow, and defoliate from mid to late summer; scattered branches die back or the entire crown gradually declines over a period of years. Foliar wilting in elms is generally believed to occur when quantities of wetwood liquid are forced under pressure into the xylem vessels and thus are carried to the crown.

The leaves first curl upward along their margins, the petioles wilt and the leaves drop off while still green or turn a dull green-brown or bronze before they fall. Slowly wilting leaves turn yellow or brown before dropping. In some cases, the leaves turn yellow or a dull green-brown between the veins and along the margins of leaves. Wilting is much more common in young trees; older trees are most likely to develop a general decline in vigour or a branch dieback in the upper crown.

In elms and poplars with wilting twigs and branches, wetwood appears as gray-brown streaks or bands in the current-season wood and often resembles wood stained by vascular wilt fungi such as Dutch elm disease or Verticillium wilt. The streaks extend from the trunk into small branches and twigs and merge to form solid brown rings in the wood of one or more seasons.

Effects on Lumber

When frozen, the wet zones in wood are more rigid than normal wood. A log with wetwood often shows no defect until the lumber sawn from it shrinks and/or cracks during kiln drying. Even when no shrinkage or cracks develop, wetwood dries only about one-third to one-half as rapidly as normal wood and thus requires twice as much energy.

Control

There is no cure or preventive treatment to avoid infection and development of bacterial wetwood. The following practices may be of some benefit.

1. The fertilization of stressed trees to stimulate vigorous growth is reported to lessen the severity of wetwood in certain individual trees. Vigorous trees, however, may be more susceptible to wetwood after fertilization than trees that are under stress.
2. Slime flux can be alleviated for cosmetic purposes by installing perforated, iron or plastic drain tubes to relieve the gas pressure and allow continual drainage away from the tree. The release of pressure probably prevents further distribution of bacteria and bacterial toxins within the tree.

   The first step is to drill a 3/8-to 1/2-inch diameter hole, 6 to 14 inches directly below the fluxing region. The hole should slant slightly upwards to allow for good drainage and extend through the heartwood to within a few inches of the bark on the opposite side of the trunk or branch. In elms, it may be necessary to drill several holes at various locations in the trunk or major limbs before the infected wood area is tapped.

The second step is to insert a piece of threaded, iron or semirigid polyethylene tubing into the drainage hole only deep enough to hold it firmly in place. How to properly drill and install the drain tubes for pruning cuts as well as crotch wounds and trunk cracks. The tubing should be long enough to carry the dripping wetwood slime clear of the trunk and root crown. A disadvantage of installing drain tubing is that another deep wound is created that breaks the "compartment" the wetwood is in, thereby allowing the discoloration and any future decay to spread outside the wetwood-affected column.

3. Remove dead and weak branches. Disinfect tools with 70 percent rubbing alcohol before pruning another tree.
4. Promptly prune and shape bark wounds. Remove diseased bark and underlying wood around infected pruning cuts and apply a disinfectant coat of shellac. An additional coating of a good wound dressing is optional, primarily for a cosmetic effect.

Chapter 6

Plant Protection

The State Plant Protection Department (hereinafter referred to as Department) is guiding the policy of state plant protection. In order to protect the territory of Lithuanian Republic from harmful organisms, diseases as well as from import and distribution of weeds, the Department performs the inspection of imported plants and researches of plant diseases, pests and weeds pervasion. The Department is responsible for the certification of plant protection means as well as their usage control. The State Plant Protection Department prepares projects in the field of plant protection and quarantine and adjust them according to the EU standards and other international agreements.

The State Plant Protection Department consists of 6 Administration Divisions, 10 Regional Plant Protection and Quarantine Centres (these centres conduct researches of harmful organisms pervasion in agricultural sown areas and participate in creating forecasts of agricultural pests occurrence as well as perform control of imported, exported plants and vegetable products. These centres are also responsible for issuance of phytosanitary certificates and provision of consultations for users of plant protection means) as well as 19 Borderline Plant Quarantine Centres (these centres are responsible for borderline control of imported, exported plants and vegetable products).

The Department employs 250 workers, about 50 employees work in the Administration, in Vilnius.

Preparation for the EU Membership

When executing plans of Law Coordination Means and Acquis Implementation Means as well as carrying out the Negotiation position covenants, all necessary acts were prepared and ratified in 2003 that were necessary for adjustment of Lithuanian Republic Phytosanitary law base according to the legal acts of the EU. During 2003, legal base of the EU was systematized; information about new legal acts of the EU in the field of phytosanitary and plant protection was collected and analysed. Tables of conformity to the EU directives were prepared for all national phytosanitary legal acts adjusted according to the EU requirements. These tables indicated the conformity of legal acts and their compatibleness with the requirements of the Union. A meeting with the officials of European Committee was summoned in order to create conformity tables of national legal acts to the EU legal acts. The experts of European Committee analysed the tables and commended them.

The order of plant protection products certification, usage, storage and trade is established in Lithuanian Republic Law on Plant Protection that shall become valid from 1 May 2004. Conformity to the requirements of mentioned Law shall be of great importance for farmers, because their allowances will depend on the proper usage of plant protection products. The European Union promotes the preparation of new member states for the membership and provides a financial support, i.e. PHARE program and bilateral partnership projects.

Quality Control and Analysis of Diseases

According to the quality requirements, in 2003 inspectors examined more than 140 000 t of imported fresh fruit and vegetables originating from 34 different countries. In 2003, Holland (onions, carrots and salad vegetables), Spain (cucumbers, tomatoes, exotic fruits), Poland (apples) were the main suppliers. More than 93 percent of this production conformed to the mandatory quality requirements. The main reasons of nonconformity were external quality defects (about 540 cases), wrong labelling or labelling absence (about 130 cases), weight (7 cases). In order to eliminate the mentioned incompliance, the retailers of fresh fruit and vegetables usually employed the method of quality class reduction (this method was applied for 2936 t) and re-assortment of products (1117 t were re-sorted) so that products would conform to the mandatory quality requirements. The similar situation was in 2002 m.

In 2003, production of Lithuanian farmers was subjected to more frequent inspections. More kinds of fruit and vegetables were inspected than in 2002. 114 cases out of 296 inspections performed showed that Lithuanian production did not conform to the quality requirements. In 2003, quality inspectors registered cases of nonconformity. The number of such cases was by 30 percent lower than in 2002. Owners of companies should not be considered as the only persons that ensured better preparation of production for sale. The Department experts that conducted seminars and provided with other information about mandatory quality requirements had also significantly contributed to it. In 2003, 2437 storage, trade and usage areas of plant protection products were inspected and as a result there were 1550 cases of major or minor violations registered.

In 2003, 2331 certificates of conformity to the quality requirements were issued.

Last year, 8478 inspections of accountancy and phonological examinations were performed. The information on researched diseases and pests is presented in the descriptions.

According to the consumed amount of plant protection products during 2003, it was established that an average of 0.25 kg of active agents was used for one hectare of agricultural landed property. In 2003, 1681.8 thousand ha of plants were subjected to the plant protection products. In 2003, 938 000 ha were subjected to herbicides in order to fight weeds (this number is by 78.9 thousand ha more than in 2002, and 419,6 thousand ha more than 1995).

Status of Protected Zones

According to the agreement of joining into the EU, Territory of Lithuanian Republic is declared as a protected area from fruit-tree bacterial blight (Erwinia amylovora) (hereinafter referred to as blight) and Beet Necrotic Yellow Vein Virus (Rhizomania) (hereinafter Rhizomania) that are spread in other EU countries as well as spread in specialised cultures that have a limited economic importance to the entire EU.

Certain plants that may carry such pests or diseases must have plant passport with a special code, consisting of letters "ZP" (i.e. zone protected), when imported into Lithuania or other protected zones. This passport must also contain codes of all countries that have a protected zone status in the EU. These codes indicate that plants conform to the EU requirements for certain pests or diseases and can be imported into the certain protected zone.

Plant Passport

From 1 January 2004, according to the ratified legal acts, the new labelling system that conforms to the phytosanitary and quality requirements of the EU became in operation, i.e. the issuance of plant passports was introduced. Plant passport is a document valid within the entire EU; it is intended to ensure conformity of plants, vegetable production and etc. to the phytosanitary requirements. So from now on, Lithuanian vegetable production shall have equal opportunities to be realized not only in the local market, but also in the EU countries.

In 2003, 84 611 phytosanitary certificates and 1022 phytosanitary register certificates were issued; 2292 phytosanitary inspections of plants and vegetable production were performed.

Performance Improvement

The employees of the Department intensively raise their qualification as well as gather information on the news in their speciality. In 2003, they participated in 20 International seminars and traineeships in the field of plant healthiness, introduction of plant passport system, bacteriological blights and Rhizomania. Seminars were summoned in the Department as well. These seminars were conducted by foreign experts that visited Lithuania 20 times last year. In 2003, the Department continued execution of the Twins project „Modernisation of phytosanitary researches and administration", dated 2001. Six foreign experts provided with consultations in the field of wood, potato growing, plant passport and protected zones. In 2003, the PHARE project „Consolidation of General Agricultural Policy" was started. One part of this project is intended for development of quality control system in Lithuania, German expert provided consultations about the issue.

The Department organized about 30 courses and seminars on various issues of quality control for inspectors working in Regional Plant Protection and Quarantine Centres as well as in Borderline Plant Quarantine Centres. The Department and Lithuanian Horticulture and Olericulture Institute trained 21 inspectors on the performance of quality checks according to the 34 fruit and vegetable positions of Composite Product nomenclature valid in the EU.

5512 users were trained by using the Curriculum for Plant Protection Means Users approved by Ministry of Agriculture, certificates of plant protection courses completion were issued that grant a right to purchase and use products intended for professional usage. By using learning program for sellers of plant protection products, 157 courses were conducted and 374 persons, responsible for trade and storage of plant protection products, were trained. During 2003, 5886 certificates of completion of plant protection courses were issued. In 2003, about 300 seminars and courses for agriculturalists were held that provided information about requirements for storage, usage and accountancy of plant protection products as well as about product control system and responsibility for violations. There were more than 60 publications on the above issues in the national press.

Electronic system Caddy for registration of applications for plant protection products was introduced. This new system will help to reduce costs and time for processing of applications.

Lithuanian Borderline

In 2003, two last Borderline Phytosanitary Control Posts in Medininkai and Lavoriškes were finished and equipped with the necessary equipment conforming to the EU requirements. The building of these posts was financed by the PHARE project (2000) and national budget. In the beginning of 2003, the quality inspectors, working in the Borderline and Regional Plant Protection and Quarantine Centres were provided with the necessary equipment for analysis of fresh fruits and vegetables. According to the legal acts, as from 1 January 2004, three kinds of Ambrosia spp. were included into the harmful parasitic plant lists. From now on, the imported grain into Lithuania shall be controlled depending from the origination country (i.e. from EU countries or not EU countries) in order to protect Lithuanian territory from importing and spreading of weeds.

Ground Application Equipment

Anyone with access only to the published records of ground spraying against tsetse would have difficulty in discovering just how the insecticides are applied. This is being remedied to some extent by the publication of more information on pesticide application equipment, a good deal of which has relevance to tsetse control. COPR and the Overseas Spraying Machinery Centre (OSMC) in collaboration with other UK scientific units through the Ministry of Overseas Development Natural Resources Advisors Coordinating Committee (ODM NRACC) Sub-Committee on Pesticide Application Overseas help in the promotion and coordination of pesticide application research and development in relation to the demands of overseas countries. Assessment of spraying machinery is ongoing research at OSMC, and many of the tests carried out are based upon recommendations by WHO. Spraying equipment is used to disseminate pesticides as liquids in the form of aqueous suspensions, emulsions and solutions. It is not possible to list all the equipment in current use or which is available, but the apparatus mentioned below has been used in major control operations; e.g. aerosol dispensers are used at fly pickets, fogging machines, although employed in the early days, have now declined in use, whereas a variety of hydraulic energy and gaseous energy equipment is still in regular service.

Hydraulic Energy Equipment

Spray Nozzles

There are three basic pattern shapes produced by hydraulic energy nozzles; the falt fan, the solid cone and the hollow cone. The limitations of the fan pattern for tsetse work are reflected in the very few occasions on which it appears to have been used. In a campaign against G. palpalis, the Eclipse sprayer was used, fitted with a lance that gave a fan-shaped spray for treating the leaves and stems of plants along paths. One difficulty which arises with these nozzles is that the width of the fan has to be orientated in order to cover the surface to be sprayed.

Cone-shaped sprays are produced by nozzles which consist essentially of a swirl plate with angled slits or holes which produce a vortex in the flow of insecticide, causing it to issue from the orifice in the form of a solid or hollow cone of spray depending on the presence or absence of a central hole in the core. Whether the solid cone nozzle has ever been used in tsetse control is uncertain.

The hollow cone spray nozzle is in general use and either a constant or variable spray width can be produced. For the former, the actual dimensions are determined by the design of the swirl core, the size of the nozzle disc orifice and the operating pressure. The practical limitations of these nozzles is that when spraying tree trunks or individual branches, the position of the nozzle, in relation to the surface being sprayed, must be constantly adjusted to ensure that a swath of insecticide of the required width is obtained.

Accurate and, therefore, less wasteful application can be achieved by a skilled operator using a spray lance (sometimes referred to as "gun") emitting a cone of spray whose shape can be instantly adjusted to fit the width of the object being treated. This can be achieved by altering the length of the swirl chamber. The angular width of the cone at a given pressure is inversely proportional to the length of the chamber between the swirl block and orifice disc. Alternatively, wastage can be avoided by adjusting the distance between the nozzle and the target surface, but this may necessitate moving the nozzle more rapidly when close to the surface.

Spray Lances

Two types of spray lances are in general use; one has a shut-off valve in the handle for controlling the flow of insecticide and produces a constant or variable cone depending upon the type of nozzle fitted. With the other, the shape of the cone is adjusted from the handle by means of a control rod inside the lance.

Shut-Off Valve Models

These consist of a trigger-operated, instantaneous shut-off valve which, with tube and nozzle, from the lance assembly. To produce a constant-shape cone, the nozzle assembly contains a swirl cone, an orifice disc and a screw-retaining cap. The shape of the cone produced for a given pressure is related to both the design of the swirl plate and the size of the orifice in the disc. Variable cone nozzles consist of a nozzle body attached to the end of the lance by means of an adaptor. Variations in the shape of the cone are produced by altering the length of the swirl chamber between the core and the disc by screwing or unscrewing the cap. Leakage of insecticide down the lance from the nozzle is prevented by an 'O' ring immediately above the threaded portion of the body.

The CP 601 spray gun has been used in some countries. This consists of a trigger-operated shut-off valve with a brass body, a plated steel trigger, a brass tube 58.5 cm in length with a 45° bend near the end and a single cone spray nozzle No. 520/P. The nozzle has a plastic swirl core, of which only one model is available, a stainless steel nozzle disc and a plastic nozzle cap. Six different orifice sizes are available, with diameters ranging from 0.79 to 2.38 mm, but in both Nigeria and Uganda disc No. 12 with a 1.19 mm orifice is fitted.

Also in this category is the Stoprex spray gun. This has a thumb trigger in contrast to the models previously described, all of which have a four-finger trigger for operating the shut-off valve. The lance and constant-shape cone nozzle assembly are, however, similar to those of the CP601, but the Rex Variable Nozzle is of a different design. Variations in swirl chamber length are made by rotating a collar to which the same stainless steel orifice disc and plastic nozzle cap as those used for hollow cones are fitted.

Control Rod Models

Spray guns of this type are intended for use in situations where frequent and often rapid changes in the width of the spray pattern are necessary, e.g. when treating the trunks of different trees and the underside of branches at varying distances from the operator. As with the variable cone nozzles described above, pattern shape is changed by altering the length of the swirl chamber. The disc is fixed but movement of the core is achieved by the control rod. The CP700 Spray Gun is one example of this model which has been extensively used in Uganda but is no longer available. A some what different mechanism for adjusting cone shape is found in the GunJet No. 12 and No. 14 spray lances whose use in Botswana has been reported. The essential difference between these and the No. 700 is that changes in the swirl chamber are made by rotating the control rod. On rotating the handle the swirl chamber is progressively opened and spray is emitted, at first in the form of a short broad cone.

As this movement is continued, the cone becomes progressively longer and narrower until a complete circle has been turned and the swirl chamber is fully open and the spray is emitted as a solid stream. The cone dimensions are related to operating pressure and orifice size and two types of fitting may be used for producing hollow cones. One is a stainless steel disc intended for the higher rates of application, while for the lower outputs the disc can be replaced by a ConeJet spray tip, of which ten sizes are available.

Another lance in which the spray cone is adjusted by means of a trigger-operated control rod has been used with a Platz vehicle-mounted sprayer in Zambia. The output of each lance is controlled by the manual operation of the trigger and can be varied between a very fine cone spray and a solid jet (37). The maximum output (from one lance) is about 4.1 per minute with a 2 mm disc orifice and 2.18 per minute with a 1.5 mm disc orifice. The maximum throw is about 15 m in still conditions.

Hand Operated Machines

In contrast to the extensive use made of hand compression sprayers, few organizations have adopted the lever-operated type, which is lighter but less robust. Unless a pressure regulating device is fitted, considerable variations in output rate can occur because of changes in lever operation, which is affected by operator fatigue. Operators have also found continuous pumping more tiring than the periodic pumping of compression sprayers and a further disadvantage is that they are unable to use both hands freely to gain access to difficult places or to steady long lances when spraying the undersides of high branches, one hand being required for pumping.

Pressure Retaining Compression Sprayers

Extensive use has been made of these machines because they maintain pressure and, consequently, less effort is required to operate them than nonpressure retaining sprayers. The high operating pressure also reduces the tendency to blockages. This type of sprayer can be rapidly charged by means of a detachable pump and maintains air pressure after the spray has been discharged; further pressure loss is prevented by a floating ball valve.

Two types of Colibri are still used in antitsetse operations, each using a different type of charge pump. The Leo-Colibri has its own detachable hand pump, whereas the Favori-Colibri machines are charged by means of a large manual or motorized pump which can charge several machines simultaneously. This can be time saving, especially where operators are working near tracks or in close proximity to each other, e.g. when treating narrow zones of riverine forest.

The detachable pump is another favourable feature of Colibri equipment because it reduces the weight and, consequently, operator fatigue, a matter of some importance in highly selective operations where accuracy is essential. In Uganda, where the Leo-Colibri is still in service, each sprayer operator is accompanied by a porter carrying both a reserve supply of insecticide and a charge pump. In Nigeria, a group of sprayers is serviced by a single mobile manual Favori-Colibri hand pump. The Colibri has been used also in the Central African Republic.

Non-pressure Retaining Sprayers

Non-pressure retaining sprayers have been widely used, often in preference to pressure retaining sprayers. This has been due to their lower cost, the nonavailability of suitable pressure retaining models and the need to stock fewer spare parts.

Some of the first insecticide trials in Uganda, were carried out with a Kent sprayer. In the past, this sprayer has seen brief service in East, West and Central Africa, but no reference has been found as to the model preferred. Three have been in production for many years with maximum working capacities of 4, 5, 9 and 13.6 1. The pump cylinder is of brass, tinned inside to prevent corrosion and is filled through a fairly small opening in the cylinder which can be closed with a plug. Another non-pressure retaining machine is the Warley, which is very similar to the Kent. This, too, is filled through an opening but sealed with a cap. In Uganda it was replaced by the Leo-Colibri before large scale spraying operations got under way, but its use in other countries is referred to by several authors. Like the Kent, there are three sizes of Warley sprayers of similar working capacities.

When the Leo-Colibri went out of production, a suitable pressure-retaining alternative could not be found. However, in both Uganda and Nigeria, the CP201 was then used, based on a Gloria machine with the same number. As with all non-pressure retaining sprayers, the pump remains in the machine while spraying is in progress. Like the Kent and Warley it was filled through an opening in the top of the cylinder provided for this purpose and closed with a plug. In its commercially available form, the CP 201 was found to be much less comfortable to carry than the Leo-Colibri. Its extra weight was a factor with the low position of the upper sling swivel support band as near the top of the tank as its curvature allowed. The commercial life of this sprayer was, however, short, and by 1971 it had been replaced by the CP 148.

The tank of the CP 148 is of stainless steel and the greater resistance to corrosion of this material may be an asset, although with proper care the corrosion of brass tanks is negligible. The tank provides greater tensile strength and the thinner sheet metal reduces its weight (7.4 kg). How well these thinner walls stand up to the falls which inevitably occur when treating vegetation on a rugged terrain remains to be seen. Unlike any of the machines referred to above, the CP 148 does not have a separate filling opening. Instead, the pump is removed and insecticide is poured in through that opening with the aid of the fairly large plastic funnel. This is firmly attached to the machine and is extended to surround and protect both the pressure gauge and the safety valve. Other models with smaller tank capacities which have been used in African countries are the Muratori, the Galeazzi and the Hudson x pert which also proved satisfactory.

A choice of either diaphragm pump or piston pump models, such as the Cosmos, is available and both have been utilized for spraying operations. There is one published record of the latter type, though these models have been used in Zambia. The Cosmos has a stainless steel 16 l capacity tank but the pump, which includes a paddle agitator, is of polyamid plastic. In contrast to other lever-operated machines, the Kestrel has a pump fitted to the outside of the tank.

A very different machine is the CP 3 which is now in service in Zambia and Uganda having replaced the Polclair. The polypropylene plastic body has a working capacity of 18 1 and mounted underneath it is a pump which is basically of the diaphragm type, but the upper diaphragm support has been so designed that it acts as a piston. The compression cylinder is not a separate component but is moulded in one piece with the pump body.

Pressure Regulators

When the fitting of a constant pressure device is to be recommended (16), an on/off tap to the outlet valve is necessary so that faults in the machine can be rectified without loss of spray liquid. Pressure regulators were once used extensively in Nigeria but were later dispensed with because spray deposit was more effectively controlled visually by the operators. A simple alternative to the pressure control valve is the disc flow regulator which has been incorporated in some sprayers.

Motorized Machines

The use of a small motorized pump is referred to in an account of spraying operations along the Malawa river on the Uganda/Kenya border in 1956. This type of pump was used from a dinghy and was capable of throwing a variable jet of spray to a distance of 8 m. It is believed that the 700 model spray guns were used in conjunction with the pump and, more recently, the banks of the same river were sprayed again from a boat. The insecticide was applied with similar guns but was supplied from a different type of motorized sprayer. This sprayer consisted of a high pressure diaphragm pump powered via a reduction gear by a 4-stroke petrol engine. Two delivery hoses were attached and an adjustable regulating valve gave a range of pressures from 10 to 30 kg/cm^2.

In Zambia the use of powered pumps has been the standard practice since large scale eradication campaigns were implemented. The equipment employed was a Platz spray unit mounted on Unimog model 411 tractor. The unit consists of a 3-stage piston pump driven by a pto shaft. Insecticide was suppled from an 800 1 tank mounted on the vehicle and was delivered through two hoses at a constant pressure of 30 kg/cm^2 to a pair of lances. These were held by two operators sitting on swivel seats at the back of the loading platform, but in rough conditions they walk alongside the vehicle. This type of equipment is, in comparison to manual methods, relatively expensive and wasteful (81) and is likely to be replaced by knapsack sprayers.

Gaseous Energy Equipment

Motorized Knapsack Mistblowers

These are used in most countries where tsetse control is practised. The model currently in use in Tanzania is the Motoblo Super 70 which closely resembles the Solo-Port 430. Both are equipped with 70 cc 2-stroke engines and 10 1 insecticide tanks.

Motoblo sprayers have been used in Tanzania since the early 1960's. Both the 60 and 90 models were also in use in Nigeria at about the same time. The model 60 was preferred to the latter because it was lighter and easier to handle. Further spraying operations with these machines have been described but they have since been replaced by the CP 40, which is now out of production. In spite of its smaller and somewhat unreliable engine, the CP 40 proved to be a reasonably effective machine. The plastic insecticide tank has a capacity of 10 1 and in Nigeria it proved effective for spraying persistent insecticides in swamp forest. A number of modifications were made to the CP 40, the most important being the replacement of the thin plastic air hose with a fabric-coated rubber hose, due to the plastic hose being very prone to perforation by thorns. This resulted in appreciable air loss and a consequent decline in the spray penetration.

Three KWH 77 machines were used in Uganda and they have been employed extensively elsewhere. Motorized knapsack mistblowers have also been employed in Rhodesia, but how long they have remained in service is not known.

Centrifugal Energy Equipment

The only record of equipment of this type used in control operations refers to the MEG. Briefly, this consists of a large centrifugal fan powered by a 420 cc air-cooled, 4-stroke engine. At the exit of the fan housing there is an aerofoil section containing a propellor which, powered by the air flow, rotates a wire cage at speeds of up to 20,000 rpm. Insecticide is fed into this cage through a nozzle and is emitted as small droplets which are carried away in the air blast.

Another type, the ULVA, is a light weight portable rotary atomizer. Insecticide is gravity fed to a nozzle in the centre of an atomizer consisting of two stacked corrugated plastic discs with serrated edges. These are rotated at speeds of up to 9000 rpm by a small 7-watt electric motor powered by a 12-volt battery. Extensive use is being made of the ULVA in crop protection work in the tropics, and in Nigeria it has been used to apply ULV formulations.

Smoke Generators

Some small scale control operations were carried out with smoke generators against G. palpalis (G. fuscipes) in Uganda (20, 47), but accurate dispersal of smoke under varying weather conditions was difficult. Efforts have been made to overcome these problems but this technique has not been used extensively.

Thermal Foggers

Field trials with the Todd Insecticidal Fog Applicator (TIFA), and its use in the treatment of small pockets of tsetse have been reported. The Swingfog has also been used, and larger versions of this machine, such as the SNIOO, are available for vehicle mounting.

Aerosol Dispensers

A problem in any tsetse control operation is the prevention of reinfestation, and an essential aspect of protective measures is the deflying of traffic by pickets. Tsetse flies which are carried by pedestrians, cyclists and private cars can be caught with nets, but this method is impractical for dealing with numerous flies which can be carried by a single lorry. In Uganda, small hand-operated sprayers are issued to pickets, and both gaseous and hydraulic energy models have been tried. The former, which are invariably cheap, insubstantial and intended for household use, are uneconomic because of their short working life and inefficient because of the wide variation of droplet sizes produced.

Good quality hydraulic energy models, such as the 2180, are more satisfactory. These are intended for indoor use but could be used in garage-type buildings erected at road pickets where lorries are treated.

Selection of Equipment

The diversity of types of spraying machines utilized in tsetse eradication work is largely attributable to the diversity of habitats occupied, not just by the genus as a whole, but by individual species. Ground spraying techniques have also been developed more or less independently in the countries where control is practised in a variety of ecological circumstances. Consequently, a variety of local factors have influenced the choice of equipment. The selection of machines is related to their spraying efficiency, cost, ease of handling, mechanical reliability, maintenance and availability of spares. Although tsetse control organizations carry out trials with different insecticides or formulations, there is rarely any mention of comparative trials of different spraying machines. Similar savanna woodland habitats of G. morsitans have been treated with spray guns and pressure-retaining compression sprayers in Uganda and by motorized knapsack mistblowers in Tanzania. In Uganda, persistent insecticides have been successfully applied in tsetse resting sites with a 700 spray gun. When treating savanna woodland and forest communities where the undergrowth is not too dense to pass on foot, or where access could be provided with a minimum of path cutting, a compression sprayer (preferably a pressure-retaining type) adapted for use with a spray gun has proved to be most practical. A pressure control valve is fitted to ensure even emission, and this is suitable for spraying trunks up to 3.5 m in height from a distance of 4 m. A much greater reach is often necessary when treating G. fuscipes infested vegetation from a boat. In such circumstances, a compression sprayer is inadequate, but a small portable, engine-driven, high-pressure diaphragm pump is used. With a motorized pump it is also possible to use an insecticide container with a much larger capacity than a compression sprayer, and this reduces refilling time to a minimum.

When dealing with thicketed or swamp forest habitats, access lanes are a costly item. Costs can be reduced by using suitable spray jets or motorized mistblowers which give better penetration in these situations.

Aerial Application Equipment

The use of insecticides has for some time been the accepted and main method of controlling tsetse flies and this is likely to remain so in the foreseeable future. Residual treatments of persistent insecticides (e.g. DDT and dieldrin) have been used both selectively and discriminatively by ground spraying methods. However, airspraying techniques have developed appreciably over the last few years, certainly to an extent where the use of aircraft to apply insecticides in antitsetse operations has progressed beyond the experimental stage. As far as the future is concerned, it seems likely that emphasis will be placed on the use of aircraft in many areas which are infested with flies, and it seems probable that any large scale control programmes will contain a significantly large serial input.

The principle reason for the choice of aircraft is the relative ease with which large scale operations can be carried out in comparison to spraying from the ground and also fewer trained staff required. Aircraft are able to apply spray to tsetse habitats which are inaccessible from the ground, and this either takes the form of placing the residual deposits in specific sites or by space spraying a wide area. Aerial applications of insecticides are likely to produce more general environmental contamination than ground spraying, except that much lower dosage rates are used, and therefore it is essential to incorporate ambient pesticide monitoring in any operational airspraying programme.

Application techniques by fixed-wing aircraft and helicopters have now reached the stage of development such that suitable equipment is available to meet most tsetse control requirements. Many large scale programmes to control tsetse flies are now based on aerial application of insecticides and, although they often involve a high financial commitment, aerial spraying may be less expensive than ground spraying in terms of unit area covered, depending on the insecticide selected and the amount of ground per unit area. This is partly accounted for by the need for fewer personnel and less logistic support in aerial operations than ground spraying and this often simplifies the operational planning. However, there are often difficulties in executing airspray operations because of the necessity to service and repair sophisticated equipment in remote areas and to adhere strictly to a schedule of sequential applications.

The Development of Airspraying Techniques

Research on application of insecticides from aircraft for tsetse control has progressed by two distinct and different methods. These methods are here referred to as "non-residual treatments" and "residual treatments". Although the use of aircraft in agriculture had its beginnings more than 50 years ago, for tsetse control, non-residual airspray treatments have been in use for almost 30 years and residual airspraying about eight years.

Non-residual treatments rely on sequential applications of very small droplets of a relatively concentrated insecticide, dispersed widely in savanna woodlands areas with the objective of applying spray droplets which will come into contact with tsetse flies, either in flight or in their resting sites . This method has been used extensively in Eastern, Central and Southern Africa. Residual spraying treatments from aircraft have been developed as an extension to the traditional ground techniques and adapted to aerial spraying of persistent insecticides to specific targets where tsetse are known to rest. These have been particularly suited to tsetse control in riverine vegetation, ecotones and drainage lines in West African conditions, especially where the tsetse habitat is restricted during the dry season.

Non-Residual Spraying Techniques

Aerial operations to control tsetse fly began shortly after the Second World War in South Africa, where Anson aircraft were fitted with extension tubes attached to their exhaust stubs. These devices were used to apply 5% solutions of DDT (102) to fly infested regions of Zululand. At about the same time in East Africa, work commenced on an experimental programme of aerial spraying to study methods of controlling tsetse fly. Work was carried out on an experimental basis and involved studies of insecticide formulations and spraying techniques with supporting physiochemical and biological assessments. Early experiments were carried out with war service aircraft fitted with crude devices, e.g. open pipes and thermal exhaust units which were used for the production of fine sprays and coarse aerosols.

Trials were ineffective in achieving control of fly but research continued throughout the years in improving application techniques, spray dispersal equipment and insecticide formulations. Eventually, improved fly control was achieved but only at high costs. The development of an improved thermal exhaust device fitted to a Cessna 182 aircraft which was capable of producing good aerosol characteristics was used to apply dieldrin and isobenzan which gave very satisfactory control of tsetse flies at reduced cost. In 1964, endosulfan was introduced for the first time and this experiment was shown to reduce costs appreciably.

These experiments pioneered the use of aircraft for the control of tsetse flies by non-residual spray methods and laid down the foundation of techniques which have developed over the last few years in Central and Southern Africa. In 1967, an airspray experiment was carried out to control tsetse fly in SW Barotseland, and this subsequently led to the commencement of a large scale spraying operation which was carried out the following year and covered an area of 1600 km^2. The techniques have been further refined in Botswana in 1973 where a twin-engine Piper Aztec aircraft fitted with a single Micronair unit has been used to control tsetse fly in the Okavango Delta.

Residual Spraying Techniques

Trials were carried out in Kenya using an experimental formulation of dieldrin applied by helicopter (Bell 470) and fixed-wing aircraft (Piper Pawnee 235) to control G. pallidipes. The insecticide was formulated as an invert emulsion and was applied from a Bi-Flon device. Spraying by helicopter proved to be prohibitive because of the high operational cost, but trials were continued using a fixed-wing aircraft.

The insecticide applied by this method was shown to penetrate the canopy and produce toxic deposits in the understorey thicket. A spraying technique was developed which consisted of applying a swath at intervals of 54 m across the thicket and this proved to be effective in controlling fly for at least eight months.

In West Africa, DDT, dieldrin and BHC as emulsion concentrate formulations have been applied primarily to clear riverine vegetation of G. tachinoides. Two applications at three weekly intervals were made from a Bell Helicopter (47G-4A 280 HP) fitted with boom and nozzle equipment. These areas have remained fly free for over four years. In 1971, discriminative spraying of dieldrin to vegetation usually sprayed from the ground was carried out in the Northern Guinea Savanna Zone (NGSZ) of Nigeria.

The area sprayed consisted of riverine vegetation, Isoberlinia woodland and ecotones around hills and in the sub-plains. A 20% dieldrin EC was applied at the rate of 1.5 to 2.0 kg/ha and, with isolated exceptions where fly pockets were missed, these areas have remained fly free for over three years. During 1972 spraying trials in the NGSZ, a helicopter fitted with boom and nozzle equipment was successful in controlling G. morsitans for several months using ULV formulations at 0.5 kg/ha and 1.5 kg/ha of dieldrin and endosulfan, respectively. Subsequently, trials using eight Ulva electrically operated rotary atomizers mounted on the boom of the helicopter indicated that tsetse fly was controlled and exterminated at 0.64 kg/ha and 0.8 kg/ha with ULV formulations of dieldrin and endosulfan respectively to about 10% of the infested area.

Equipment for Non-residual Insecticide Treatments

Open Pipe

The first trials carried out in East Africa employed an Avro-Anson XIX aircraft fitted with four spray tanks having a total capacity of about 800 l. Spray liquid was emitted under gravity from two wide pipes which extended about 35 cm below the fuselage of the aircraft. Emission rate could be adjusted by an Iris diaphragm which was situated at the end of each pipe. Adjustments were made by the pilot on the cockpit during the flight and spray atomization was achieved by the action of the slip stream which broke up the jet of liquid into spray droplets as it was emitted from the pipe. It was not surprising that the droplet spectrum from this rather crude device was fairly wide and the unit was subsequently discontinued.

Thermal Exhaust

One of the first devices to be used for the production of aerosol droplets was fitted to an Avro-Anson XIX aircraft. The coarse aerosol was produced by allowing the solution to flow into a modified exhaust system. The solution flowed through a narrow pipe 22 mm internal diameter and into the exhaust pipe at a distance of 50 cm from the centre line of the exhaust manifold. It then passed down along the exhaust tubes 10 cm in diameter, and was finally emitted vertically downwards into the slip stream about 30 cm below the trailing edge of the wing. The installation was based upon a design used for similar work against tsetse flies in South Africa.

In 1962, research was reopened to develop thermal exhaust units for aerosol production to enable studies to continue on aerial applications of insecticides for tsetse control. A device was developed for a Cessna 182, and this consisted of an Inconel tube 45.7 cm in length and 74.6 mm in diameter which was secured to the exhaust stub of the aircraft. The spray liquid was fed into the exhaust extension through a metering jet located outside the exhaust system. Another device consisted of fitting extensions to twin exhaust pipes, the extensions being two slightly curved stainless steel pipes 5.1 cm in diameter and approximately 60 cm in length.

A 340 l Sorensen spray tank was fitted below the fuselage of the aircraft and the system incorporated a 3-bladed, fan-driven Simplex centrifugal pump (2.54 cm). The flow of liquid was controlled by two diaphragm check valves, each fitted with a strainer and a D4 orifice disc. The liquid then flowed through two copper tubes 0.32 cm internal diameter to the outlet ports of the extension pipes.

The point of entry into the extension was 20 cm below the manifold where the temperature was estimated to be 550°C. Thermal exhaust units are currently in use in Zambia with the parastatal organization, "Rural Air Services", which has three Beechcraft "Baron" twin-engine aircraft with Sorensen external tanks of 360 l capacity.

Rotary Atomizers (Air Driven)

Micronair has produced a number of rotary atomizers over the last few years and these atomizers have been used on a variety of aircraft in numerous pest control operations throughout the world.

The physical characteristics of the spray emitted from the rotary atomizers are determined by the liquid throughput and the speed of rotation, but the droplet spectrum achieved is more uniform than hydraulic spray nozzles. The units are driven by a 6-bladed wind mill, whose blade angle and, therefore, speed of rotation can be adjusted between flights. Micronair equipment is used mainly for the ULV application of mists for the control of insects and diseases in crop protection and can be modified to produce aerosols in the range of 10–40 microns in diameter applied from small single-engine aircraft, provided that the insecticide solution is fairly volatile.

The latest unit to be manufactured in this series is the AU3000 Micronair rotary atomizer, and this has recently undergone trials in Botswana fitted to twin-engine aircraft to control tsetse fly. This has a smaller diameter cage (12.7 cm) than previous models, and when fitted to a Piper Aztec spraying at a speed of 260 km/h gave droplets of approximately 25–30ì vmd using insecticide solutions containing diesoline and Shellsol AB.

Equipment for Residual Insecticidal Treatments

Bi-Flon Apparatus

The Bi-Flon apparatus was designed specifically for the application of pesticides formulated as invert emulsions. Invert emulsions are produced when invert oils are mixed in water, the oil phase becomes continuous and the water phase dispersed.

The greater ratio of water to oil, the more viscous the emulsion. The use of inverts was designed to develop drift-free spraying, but the equipment and the method proved to be too complicated and the practice has now been discontinued. However, it is worthwhile mentioning some experiments which were carried out to control G. pallidipes and G. brevipalpis in the Lambwe Valley, Kenya because it may well have a bearing on future spray programmes using residual insecticide formulations.

Trials with a Bell 47G helicopter proved to be very effective in producing appreciable reductions in fly numbers, but the technique was too expensive and investigations were not pursued further. Trials were continued with fixed-wing aircraft and efforts to cut the cost by reducing both insecticide and spraying time were successful. Studies were made to evaluate the effect of spraying at intervals of 54 m, and it was shown that at least 60% of the area would have to be sprayed to achieve any satisfactory degree of fly control.

Boom and Nozzle Equipment

Residual spraying has been carried out in West Africa where helicopters have been in use to apply persistent deposits from the air in the Sudan savanna zone in Niger in 1969. An area mainly infested with G. tachinoides in dense riverine forest was sprayed twice with mixtures of DDT, dieldrin and BHC in an emulsion concentrate using a Bell helicopter 47G-4A fitted with boom and nozzle equipment.

In 1971, a trial covering 240 km^2 in the NGSZ in Nigeria was carried out with the same helicopter fitted with two spray tanks, each of 100 l capacity fixed to the right and left sides of the engine. A rotary pump which was belt driven from the engine was employed with 32 D3/25 nozzles at intervals of 50 cm. Applications were made at speeds of 32–40 km/h spraying to within 1–2 m of the canopy, and this gave an effective swath of about 20 m. In larger broken areas of Isoberlinia woodland, swaths were sprayed at intervals of 200 m. Only about 10% of the area was treated according to the discriminative ground spraying technique developed earlier for ground spraying. Dosage rates were reduced and techniques improved, and this was achieved by using fewer nozzles on the boom, thereby reducing the liquid throughput. In 1972, 12 of these nozzles were used to apply a ULV formulation of endosulfan in a vegetable oil. This ULV technique enabled helicopter applications to be made more efficiently in terms of insecticide used and in the overall cost of the operation.

Rotary Atomizers (Electrically Powered)

During the period 1972–73, spraying techniques in Northern Nigeria were further refined by employing modified Ulva atomizers and eight units were fitted to a Bell 47G helicopter. Three were fitted on each side of the boom and two behind the engine. Each atomizer was driven by a 24-volt motor and rotated at a speed of 9000 rpm. Liquid was carried from the boom through a restrictor to the atomizers and a constant flow was achieved. The system allowed for multiple and single operation of the atomizers during the flight, and the spray techniques employed were similar to those used in earlier trials. The helicopter flew at 25–32 km/h over dense riverine vegetation and 40–47 km/h over open Isoberlinia woodland where spray swaths were made at intervals of 200 m.

In 1973, an investigation was carried out to study the penetration of spray droplets when applied from electrically operated rotary atomizers fitted to a helicopter to a forest habitat in Ivory Coast. A Bell 47G-4A HP was used for the trials and carried two tanks, each

100 l, fixed to right and left sides of the engine. The rotary pump, driven by a belt from the engine, supplied the spray system with liquid. The system consisted of eight electrically driven prototype Ulva atomizers which could be individually activated. These were attached at intervals of 0.5 m along a centre boom and two outer booms having a combined span of 10.5 m. The 24V/60W atomizer motors gave a nominal rotation speed of 9000 rpm unloaded. Spray droplet sizes depended on throughput and this could be altered by changing the size and number of the liquid feed nozzles.

Experiments have also been carried out to study the effectiveness of several ULV insecticides applied by helicopters to control G. palpalis gambiensis and G. tachinoides in the area of the WHO Onchocerciasis Control Programme in the Volta River Basin. The helicopter used was a Bell 47G-4A, 305 HP with a speed range of 32–80 km/h, and two rotary atomizers secured 6–8 m apart were used to apply ULV insecticide formulations. Two types were used: one type produced droplets between 20 and 30ì in diameter which were applied at a rate of 0.6 l/min; the other type, which had large apertures, produced droplets of 80–100ì diameter at 1.2 l/min.

The helicopter flew at a constant height of about 2 m above the canopy, but it was necessary for it to make side turns to follow the bends in the water course. The powerful air current propelled downwards and backwards by the rotor carried the insecticide, which assisted the penetration of spray into the vegetation at ground level. In order to avoid losses due to convection near the surface of the ground, ULV applications were conducted for about two hours after sunrise.

Types of Aircraft Suitable for Antitsetse Operations

Small single-engine, fixed-wing aircraft are of limited use in tsetse control operations except possibly in the carrying out of experiments where relatively small areas need to be treated. In most instances, aircraft below 2500 kg maximum take-off weight would not have the range or weight carrying capability required for such programmes. It is probable that no single type of aircraft would be suitable for large scale tsetse control programmes and especially if a range of different tsetse habitats require treating.

Large flat areas can be satisfactorily sprayed with medium and possibly large multiengine aircraft, while small mountainous areas, often found in Eastern Africa, are probably more suitably treated by light twin-or medium single-engine aircraft. Mountainous areas, areas difficult of access and especially winding riverine habitats are treated more effectively by a light helicopter. In general, the smaller types of fixed-wing aircraft used in tsetse control operations should have good pay load and range characteristics, be easy to control at low flying speeds and have good short take-off and landing (STOL) characteristics.

In addition, they should allow the pilot good forward and downward visibility and be corrosion proof against the effects of insecticide chemicals. Tsetse control operations are often carried out in remote wooded areas where the opportunities for making a safe emergency landing (e.g. in the event of engine failure) are limited. For this reason, multiengine aircraft should be used whenever possible. Twin-engine aircraft should have a satisfactory single-engine performance at maximum take-off weight up to 2500 m density altitude.

With helicopters, important considerations are the forward and downward visibility for the pilot and the manoeuverability of the aircraft. The helicopter may be required to land in unprepared areas and a shielded tail rotor and high lift skids are valuable features. Any type of helicopter can be used in tsetse control operations, although machines with a maximum takeoff weight in excess of 1500 kg will probably be prohibitively expensive to operate.

Despite rising prices which have recently affected all aspects of pest control throughout the world, airspraying costs for tsetse control have actually decreased since techniques were first developed. This is attributed to a number of things but, primarily, spraying equipment design, better insecticides and improved application techniques have all played a major role in reducing costs. Improvements in dispersal equipment have resulted in more efficient spray atomization which, as a consequence, has helped develop application techniques. Aircraft utilization has increased as a result of extending swath widths and high spraying speeds. Costs have also been reduced because of better pesticides which have been discovered and formulated as concentrates in aromatic oils containing higher percentages of active ingredients than was possible when field trials were initiated.

At periods of peak demand, the work output of aircraft can be extended by the employment of additional pilots over a seven day working week or by fitting high powered lights in the nose and on the wings of the aircraft which makes it possible to fly at night. Typical work output data have been derived for various types of aircraft which are, or may be, used in tsetse control programmes.

Navigation Aids

No ground markers are needed for helicopters in West Africa when treating tsetse in areas such as clearly visible drainage lines and vegetation interzones. However, traditional methods of swath marking for aerial spraying in tsetse control operations over large woodland areas have been in use for some time, e.g. smoke fires, flags, balloons, marker bands, signal flares or lights. The development of tsetse control techniques using fast aircraft, flying longer spray runs, wider swaths and night flying has made the development of more sophisticated navigational aids desirable. Several types of airborne guidance systems are under investigation:

One system is based on an airborne phase measuring device which measures the phase difference between the radio emissions from two or more ground stations. The equipment is essentially a refinement of the Decca Navigation aircraft guidance system and has the disadvantage that ground transmitting stations are required sited at accurately mapped locations. Although easily portable ground stations are now available, these may be difficult to site correctly in remote areas.

Another system consists of a sensor unit which measures the inertia changes induced by altitude and acceleration changes in the aircraft during flight. The inertial navigation systems (INS) are entirely self-contained within the aircraft and no ground stations are required. The main disadvantage of INS in vector control operations using light aircraft is the cost and weight of the equipment and the fact that the system is difficult to adjust to flying reciprocal tracks.

Medium and larger types of aircraft can be fitted with highly accurate electronic navigation equipment (e.g. Doppler, VLF), but ground markers are required to line up on approach and exit from large blocks, especially when spray runs are many kilometres in length. In addition, these aircraft may be equipped with instruments to allow flying at night within close proximity to the woodland canopy. As weather conditions at night are often suitable for insecticide applications from aircraft, nocturnal tsetse control operations have been studied.

Chapter

7

Cultural Control Practices

Cultural Controls: Making Your Site Unattractive to Pests

Cultural control methods include properly selecting and rotating crops, sanitizing and solarizing the soil, choosing the best planting and harvest times, using resistant varieties and certified plants, taking advantage of allelopathy, and intercrop-ping.

Crop Rotation

Certain pests are more common in some crops than in others. Rotating crops to different sites can isolate pests form their food source or can change the conditions pests must tolerate. If another site is not available, change the type of crops grown in the garden plot. Do not put members of the same plant family in the same location in consecutive seasons. For example, do not follow melons with cucumbers or squash. This is also true for rotations using green manure crops, which add organic matter to the soil when they are tiled in before they produce flowers or seeds. Waiting two years to plant the same family of vegetable in the same location is the most effective rotation practice; however, yearly rotations can also be beneficial. Rotating annual flower plantings is also a good practice.

Sanitation

Many organisms responsible for disease and insect problems overwinter in plant debris such as shriveled fruit. Diseases on these shriveled fruit infect new leaves following spring. Removing crop residues, weeds, thatch, and volunteer.

Soil Solarization

A clear plastic sheet spread over the soil traps solar heat, which kills soilborne diseases, insects, nematodes, and many weed seeds. The treatment should occur during summer's high air temperatures and intense solar radiation. Keep the soil damp during the solarization process, and keep the plastic in place for several weeks. OSU Extension Fact Sheet EPP-7640 explains soil solarization in more detail.

Timed Plantings and Harvests

Many crops may be planted or harvested early to miss heavy pest infestations, while still achieving a full yield. Plant-ing earlier than normal may involve the use of cold frames or hot caps to protect seedlings form the weather while they get a head start growing. The crop then has a competitive edge over pests. Early planting depends upon the gardener knowing the emergence times and life cycles of the pests to be controlled.

Resistant Varieties

When buying seeds or plants, try to choose those with built-in resistance to diseases and nematodes. Sources for this information include OSU Extension Fact Sheets, seed catalogs, and plant and seed packages. It may be better to forego some production capability in favor of the increased pest resistance, if you must make such a choice. During the growing season, stressed plants can lose their resistance to pests, so be sure the crop has the water and nutrients it needs. When shopping for seeds and plants, check the labels for abbreviations similar to these, used to designate various types of pest resistance or tolerance:

A—Alternaria stem canker

ALS—angular leaf spot

ANTH—anthracnose

CMV—cucumber mosaic virus

DM—downey mildew

F—Fusarium (race 1)

FF—Fusarium (races 1 & 2)

MDM—maize dwarf mosaic

N—nematode

NCLB—northern corn leaf blight

PM—powdery mildew

SCLB—southern corn leaf blight

St—Stemphylium (gray leaf spot)

SW—Stewart's wilt

TLS—target leaf spot

TMV—tobacco mosaic virus

V—Verticillium.

Certified Plants

When they are available, consider buying plants labelled as "certified" or grown and inspected under sterile or quarantined conditions. Certified plants may cost more than others, but the certification guarantees they are free of diseases. Strawberries and potatoes are among crops which may be offered as certified plants.

Allelopathy

Allelopathy, a natural chemical interaction among plants, has been the subject of much research recently. Allelopathy refers to stimulatory as well as inhibitory properties. A living plant may release toxins, or in the case of decaying plant tissues, microorganisms may play a role in the release of the toxin. The microbes may also modify nontoxic compounds into toxic compounds. Black walnut trees and Johnson grass are among plants that have been shown to inhibit the growth of winter annual weeds and may offer some control of root knot nematode.

Intercropping or "Companion Planting"

The premise of companion planting is that certain plants repel insects, or attract beneficials that attack the insects. There is no significant data to prove the value of companion planting or intercropping, but it is thought that certain plants may produce substances which confuse insects, altering their impact as a pest. Some evidence also shows that planting flowers among vegetables attracts beneficial wasps seeking the flowers' nectar, and those wasps lay their eggs in the larva of certain pest species.

There is a popular but largely inaccurate belief among gardeners that marigold will control nematodes and other insects if planted among vegetables. Most marigold varieties do not have this capability. Only the French marigold (*Tagetes patula*) varieties, such as Nemagold.

Vegetables Families

Tomato Family: tomato, potato, pepper, eggplant.

Onion Family: onion, shallot, leek, chive, garlic.

Beet Family: beet, Swiss chard, spinach.

Cole Crop Family: cabbage, cauliflower, broccoli, brussels sprouts, bok choy, collards, kale, kohlrabi, mustard, radish, rutabaga, turnip.

Legume Family: bean, pea, cowpea, peanut.

Carrot Family: carrot, celery, celeriac, parsley.

Cucurbit Family: cucumber, watermelon, cantaloupe, pumpkin, squash, gourds.

Lettuce Family: lettuce, chicory, endive.

Green Manure Crop Family: hybrid sudangrass, buckwheat, soybean, cowpea, mung bean, garden pea, fava bean, ryegrass, rye grain, barley, oats, vetch, Austrian winter pea, clovers, greens.

Cultural Control of Drinking by Universal Moderation or Differential Access

The prospects for a new generation of cross-cultural hypotheses and analyses of drinking behaviour quietly improved a good deal over the past decade because of the emergence of a new and still largely untapped store of data. These data come from a growing series of broadly comparable social surveys focused on drinking behaviour and carried out in a variety of cultural settings. The trend began in 1970-1979 with, first, a four-country Scandinavian drinking survey carried out in Iceland, Norway, Sweden, and Finland (Makela, 1984, 1986) and, second, a three-culture study coordinated by the World Health Organization and carried out in community-level samples in Mexico, Scotland, and Zambia (Rootman, 1983). The WHO study — called the "WHO Community Response to Alcohol-Related Problems" project — spawned two additional comparable and nearly contemporaneous surveys in Canada (Smart, 1980) and in the U.S. (Roizen, 1981). More recently, surveys incorporating at least part of the WHO questionnaire have been done in Spain (Caetano and Martinez, 1987), in Japan (Clark, 1986), in a second region of Mexico (Caetano and Medina-Mora, 1986), and among Spanish-surname subgroups in the U.S. general population (Caetano, 1986). An August 1986 conference in Washington, D.C. — which drew participants from researchers involved in the original WHO project, from researchers at work on subsequent

WHO-like projects, and from others interested in this emerging resource—bespeaks the continuing vitality of this new tradition of research.

These studies offer a tonic to cross-cultural research not only because they are new and unfamiliar but because survey data bring a new kind of information to the territory, thus forcing new modes of thinking upon us. And although these survey data are not as richly textured as ethnographic reports nor as comprehensive in their coverage of the world's cultures as the Human Relations Area Files, they nevertheless offer the field a valuable new window on cross-cultural variation and a methodology that nicely complements the more familiar brands of cross-cultural data. For example, survey data almost invariably require an observer/analyst interested in cross-cultural variation to pay equal heed to variation *within* each cultural setting as well. Cross-cultural hypotheses trying to make their way in this sort of data environment must somehow manage to incorporate both cross-cultural and within-cultural variation into their explanatory agenda — no small duty.

Today, I would like to illustrate some of the charms of this sort of data by drawing on WHO-study data from the three original settings and the U.S. data, which in fact derive from a survey of Contra Costa County California in 1979. I have two goals. One is to illustrate how ordinary survey data can be used to examine structural and normative issues, issues that are usually reserved for other kinds of data and analysis styles. The other is to offer one or two conjectures on cross-cultural theory borrowing from these same data. Before we begin, however, a word or two should be said about the data sources. My survey data comes from seven separate samples: one in California, one representing the Lothian (Edinburgh) region of Scotland, two representing two kinds of communities — one urban and the other rural — in Mexico, and, finally, three samples representing an urban residential, periurban, and rural settlement in Zambia. Technically speaking, it is problematic to combine the samples drawn within a given country into a single sample, but that has been done nevertheless. Similarly, you should be aware that these data do not reflect the national populations of these settings but merely specific communities therein. Data, moreover, have been weighted everywhere besides the U.S. in order to bring the samples into equality across the genders.

Cultural Permission to Drink

Who respondents in Scotland, Mexico, Zambia, and the U.S. were asked to estimate much drinking would be "all right" for males and for females of four specific, hypothetical or test ages, 16, 21, 40, and 60. Four pre-coded answers were offered: (1) no drinking, (2) one or two drinks only, (3) enough to feel the effects but not drunk, and (4) "getting drunk is sometimes alright." The data I will be presenting are the frequency distributions for the whole samples in each setting; they tell us not the answers of males or of females regarding these age/sex statuses but instead the answers of the whole population about males and about females.

To begin, let us consider the normative patterns relating to the cultural control of access to drinking in general, a question we can address by initially collapsing together all three pro-drinking response categories. The Scottish and U.S. "permission" patterns were very similar. Both showed little difference in permission rates between the sexes (though females ran slightly below males), and both suggested an age-plateau form with high— and even nearly universal — permission granted for the three test ages twenty-one and over. Simple as they are, these curves seem to hint at three kinds of connections between drinking norms and the wider cultural environment — these links appearing as cultural premiums (1) on majority age, (2) on universalism or egalitarianism, and (3) on legalism. Concerning majority age, the normative pattern we see is consistent with a high cultural premium on a single, consensual, and officially-sanctioned "threshold" age of majority, one governing the right to drink as well as other adult prerogatives. The flatness of the curves across the three "adult" ages and the similarity of the curves for the sexes suggests a high cultural premium on universalism or egalitarianism. Finally, I see legalism hinted in the close agreement between popular opinion and official laws governing the right to drink. Legalism, in turn, suggests that the state holds and articulates the issue of cultural approval for drinking in the Scottish and U.S. settings.

The Mexican and Zambian permission patterns were quite different both from each other and from the universalistic and age-plateaued pattern of the Scottish/U.S. data. The Mexican data showed a substantial divide in cultural permission across the sexes. The Zambian data, on the other hand, showed wide variation in permission across test ages. Notice, for example, that all four female age-statuses in the Mexican sample are accorded lower rates of cultural permission than three of the four male age-statuses.

We have ordered the Zambian results from highest to lowest cultural permission level (first 40-year-olds, next 60-, next 21-, and, finally, 16-year-olds). Notice in this chart that the results follow a clear-cut pattern of age stratification: 40-year-old males and females ranked 1st and 2nd respectively in cultural permission levels; 60-year-old males and females ranked 3rd and 4th; 21-year-old males and females ranked 5th and 6th; and, finally, 16-year-old males and females ranked next-to-last and last. To be sure, no matter what the age-status, Zambian males were accorded higher levels of permission than females (even considerably higher for 21-year-olds), but more importantly, females in each age-status group were consistently accorded a permission level at least as high as that accorded males in the next-lower test-age group. Permission levels in the Zambian case, then, can fairly be said to be stratified primarily by age whereas permission in the Mexican case was stratified primarily by sex. In both the Mexican and the Zambian cases, moreover, permission to drink appears to involve a *more gradual or graded* cultural license in place of the *sharply divided* pattern at majority-age suggestion in the Scottish and U.S. permission norms.

The Mexican and Zambian normative patterns suggested neither the majority-age premium, nor the universalism, nor the legalism hinted in the Scottish/U.S. pattern. On the contrary, in both Mexico and Zambia the data reveal that the right to drink is much more strongly tied to one's ascribed status, or, in other words, to bases of social stratification more characteristic of traditional societies. Mexican and Zambian permission norms, then, seem to have 'stained-out' (to use Duster's [1983] evocative term) the ascribed-status bases of each society's broader social stratification systems, much as a biologist's dye stains tissue on a microscope slide. Mexico's gender-based and Zambia's age-based stratification systems are not in and of themselves particularly surprising. Mexico, of course, is well known for its machismo men and modest women, as Africa is known as home to many age-based systems of social stratification. Yet the striking difference between the two bases of stratification — gender and age — raise interesting questions. What are the chief differences between ascription based on age and on gender (La Fontaine, 1978)? How is it that permission-to-drink norms reflect the underlying architecture ascribed stratification in each cultural setting? What does the apparent parallelism between drinking permission and stratification tell us, on the one hand, about drinking and, on the other, about stratification? What are the chief differences in drinking practices, attitudes, and problems between universalistic and ascriptive social structures?

Finer-grain puzzles are also suggested in these findings. What does it mean, for example, that in Mexico *21-year-olds* (of either sex) were granted higher permission levels than like-sexed *60-year-olds*, whereas in Zambia the reverse was true — *60-year-olds* were granted higher levels 21-year-olds? Did the gradual, non-plateaued form of Mexican and Zambian permission norms imply that adulthood itself is likewise defined in an incremental or cumulative way? Why did the norms applying to 60-year-olds decline in levels of permission from those of 40-year-olds in Mexico and in Zambia as well as (if only in a hinted-at form) in the Scottish and U.S. data?

Permission to Drink and the Power of the State

If the permission-to-drink normative patterns hinted at different levels of involvement by the state in the regulation of drinking, then just such a relationship was unexpectedly lent support elsewhere in the WHO data, once more nicely illustrating the potential for survey data to address structural societal variables. As the study's title suggests, the WHO researchers were interested in community responses to alcohol-related problems. Its survey questions in this territory happened to include a series of four vignettes about hypothetical alcohol-related dilemmas. This series provided data on the perceived severity of alcohol-related problems in the community, on perceptions of such problems' perceived prevalences, and — most relevant here—on preferred kinds of social control responses. (Data on only three of the four vignettes will be considered here.) The first vignette described a man who had drunk so much that "he falls down on the road and cannot get up" (MANFALL); the second described a man who has on two occasions struck his wife when he was drunk (HITWIFE); the third described a man "who spends so much money on drinking that there is not enough food for his family" (NOFOOD).

Included in the battery of questions asked about each vignette were three questions asking respondents to evaluate the appropriateness of the involvement of various people or agencies in helping or controlling the man in question. For example, the questions asked (1) whether the man's relatives ought to help him, (2) whether a bystander on the street should do something, and (3) whether the police or appropriate authorities should be called in.

This series was introduced into the survey questionnaire in order to monitor norms and attitudes regarding the social handling of one or another sort of drinking-related problem. Yet its results also painted a clear picture of the relative ascendancy of the differential involvement of the state — represented by the "police/authorities" involvement — across the Scottish, Mexican, and Zambian samples.

Thus, it is worth briefly examining the different Patterns of control reported in these three setting. In the Zambian samples roughly 90 percent agreed that relatives should help in all three vignette situations. Concerning bystander involvement, 90 percent supported bystander involvement for the MANFALL situation and from 75-80 percent supported it for the HITWIFE and NOFOOD situations. Less than half the samples supported the involvement of the police or appropriate authorities in the MANFALL and HITWIFE situations, but almost three-quarters supported such involvement in the NOFOOD situation. Involvement by police/authorities ranked third — behind relatives and bystanders — for all three situations. The Scottish results were quite different. There, the vignette situations were seen as highly differentiated and sorted into two groups — the MANFALL situation, on the one hand, and the HITWIFE and NOFOOD situations, on the other. For all situations, involvement by relatives was supported by 80-85 percent of the sample. A like percent supported bystander intervention in the MANFALL situation, but involvement by police/ authorities in this situation was supported by less than half the sample. The HITWIFE and NOFOOD situations — both of which involve third-party victims of the central actors drinking — were not seen as appropriate occasions for the involvement of bystanders by roughly three-quarters of the sample; in these, police/authorities was approved by over three-quarters of the sample. The patterns suggest an alternativity relationship between bystander and police/authorities involvements and a clear division of responsibility between the benevolent interventions (of a bystander) and the potentially punitive interventions (of police/authorities) — with state involvement contraindicating involvement by the citizen on the street. Clearly, the state's monopoly of police power and the differentiation and definition of its role is much greater in the Scottish than in the Zambian case. Interestingly, the Mexican pattern of responses appeared to fall between the Scottish and Zambian patterns.

Charts 14D, E, & F present the same data as 14A, B, & C this time contrasting each cultural setting with respect to a single vignette situation. Chart 14D reports the MANFALL situation, in which all three cultural settings reveal quite similar response patterns — high on relative and bystander involvement and low on police/authorities involvement. In the HITWIFE situation, however, Chart 14E throws the issue of state control in the Scottish setting into sharp relief. Here we see that the Zambian response pattern to HITWIFE is not much different from the response pattern to MANFALL. The Scottish pattern shows a clear preference for the involvement of police/authorities and only low support for bystander involvement. In the Mexican samples bystander and police/authorities involvements are roughly equally supported by about two-thirds of the sample. Finally, in the NOFOOD situation, the Zambian pattern shows more support for police/ authorities involvement, but such involvement appears to be added to high support levels for involvement by bystanders rather than substituting for it. In the Scottish and (though less so) in the Mexican case, as noted already, involvement by police/authorities is perceived more as an alternative to bystander involvement.

These patterns, then, give us a clear if only cursory glimpse of the normative expectations surrounding relatives, citizens on the street, and civil authorities or police in regard to exemplary problematic situations. They illustrate the marked difference in the expectations regarding state control across the three cultural settings examined and lend a measure of support to the idea that the growth of the state's power is probably structurally associated with the universalistic normative patterns evidenced in the earlier permission-norm data we examined.

Cultural Permission To Drink Enough to Feel Alcohol's Effects

Though the Scottish and U.S. norms governing access to drinking suggested a universalist form, it is well known, of course, that even in these settings drinking behaviour is strongly associated with (or marked by) gender and age variables. Indeed, the WHO data from this same study reaffirmed this long-standing truth. For example, fully 80-85 percent of Scottish and U.S. young males (18-29) reported drinking at least as often as once a week; whereas among older females (50+) the proportion drinking that frequently dropped to only 25-35 percent. Concerning such relationships, the draft WHO monograph noted that "when we examine frequent drinking, in all four cultural settings the distributions for males of all ages fall entirely above the distributions of their female compatriots." What light might the normative data shed the normative arrangements for heavier drinking? Let us return data from the age/sex drinking norms series, this time with that series' responses dichotomized so as to divide respondents who would permit "feeling the effects" or sometimes getting drunk" to a hypothetical status from those who suggested "no drinking" or only one or two drinks maximum. This dichotomy will tell us more about the normative structuring of higher-quantity drinking in each setting.

The Scottish and American patterns are interesting both for their similarities and their differences. First, it is apparent in both that permission to drink is not equivalent to permission to drink enough to feel alcohol's effects. In both sites, too, the permission distributions now taken on more shape — that is to say, deviated from the "flat plateau" age-structure evidenced in the earlier permission-to-drink curves. This suggests more of a culturally sanctioned age-based hierarchy with respect to higher-quantity drinking. In both settings and for both sexes, moreover, the apparent age-hierarchies evidenced in this new shape are the same: first, 40-year-olds, next 21-year-olds, next 60-year-olds, and, finally, 16-year-olds — this, a status with very low approval for higher-consumption drinking. Interestingly, the age-based hierarchies for both Scottish and U.S. females reveal a wider gap in permission levels between 40-year-olds and 21-year-olds than occurs between 21-year-olds and 60-year-olds. The same relative spacing is in evidence for Scottish males, if more weakly, but disappears for U.S. males.

Perhaps the most striking result of the Scottish and U.S. comparison, however, is the higher permission levels enjoyed by Scottish males in contrast with either Scottish females, U.S. males, or U.S. females. To take a More than 60 percent of the Scottish sample approved feeling alcohol's effects for a 40-year-old male, whereas the comparable approval levels for 40-year-old Scottish females was a little less than 50 percent, and for U.S. 40-year-old males and females roughly 40 percent. These results imply that the hierarchy of higher-consumption permission in the Scottish sample is ordered both by sex and by age whereas in the U.S. data — though small gender differences permission rates are evident — the permission hierarchy is geared to age alone.

The Mexican and Zambian data reveal different patterns. In the Mexican sites, permission to feel the effects is very low for females of all four ages. Indeed, females are by and large simply not permitted to drink enough to feel alcohol's effects — the highest approval level was won by 40-year-old females and that amounted to only about 15 percent. For approval ran from about 30 percent, for 21-and 60-year-olds to as much as 45 percent for 40-year-old males. Notice that although 21-year-old males received a ten-percentage point higher approval level than 60-year-olds for drinking, per se, approval levels for feeling the effects were roughly equal for the two ages. Sixteen-year-old males received approval from only about 5 percent of the sample, a figure that appears to be quite low until it is contrasted with the approval levels granted females. Zambia's pattern is different yet again. The striking finding here is the high levels of approval granted "feeling the effects" for 40- and 60-year-old males and females. On balance, the data for Mexico, Scotland, and the U.S. have shown that approval levels for feeling the effects among the three statuses 21-years or older run about half as high as approval levels for any drinking — a little higher for Scots, a little lower for the U.S. and Mexican samples. In the Zambian case, however, the ratio of approving effects to approving drinking jumps to roughly 70 percent. In the Zambian case, then, approval of drinking is almost equivalent to approval of feeling alcohol's effects — suggesting the possibility of quite different structural arrangements for drinking.

Some Speculations

It is intriguing to contrast the U.S. and the Zambian patterns with respect to permission-to-drink and higher-consumption permission. The U.S. norms suggested, of course, a pattern that permitted drinking universally to all adult statuses but imposed a normative obligation to restrict drinking to moderate levels. In the Zambian case permission to drink is a differentially allocated but the norms on consumption are much weaker. I have called these two models of normative control "universal moderation" (UM) and "differential access" (DA) — and it will be noted that they provided the title of my presentation today. My hunch is that these two models of normative control may carry broad implications for the structuring of drinking behaviour and offer an interesting line of connection between drinking behaviour, on the one hand, and other dimensions or aspects of culture.

The responses we have examined refer to "ideal" norms rather than actual conduct. We have seen, however, that even ideal norms in UM normative settings like the U.S. and Scotland pertain more to permission to drink per se than to heavier drinking. Drinking per se, then, appears to be symbolically associated with the universalistically defined rights and responsibilities of adulthood in these settings, but heavier drinking is symbolically associated with a separate stratum of cultural meaning, and one drawing on more traditional status arrangements.

I was particularly impress, for example, by the close agreement in heavier-consumption norms between U.S. and Mexican males. This result may, of course, be pure coincidence, but it nevertheless highlights the interesting possibility that drinking norms, particularly in less traditional cultures, occupy a point of tension or stress between two competing normative systems.

The difference between the U.S. and Scottish heavier-consumption norms is also interesting. If the structure of drinking's norms were tied firmly to the level of modernization in a society, then we would generally expect Scottish and U.S. norms at both permission levels to be the same. The standard indicators of modernization theory are, after all, quite broad and would give us no reason, I suspect, to call either the U.S. or Scotland the more modernized of the two. It follows, then, that the observed differences in U.S. and Scottish heavier-consumption norms suggests that modernization-type variables are not the only dimension of culture at work in shaping drinking's normative structure.

This, I think, is a telling possibility and may open up an interesting analytical approach to teasing out of survey data which aspects of culture, after all, are tied most securely to drinking. Just such findings, as it happens, can prove helpful in interpreting some of the other data collected in such surveys.

For example, indicators of "alcohol dependence" in the Scottish and U.S. samples revealed strikingly similar frequencies in the two samples across six of the seven items used. The seventh item, however, showed U.S. males and females reporting much higher frequencies than the Scots. This item, as it happens, asked if one had felt that he or she should stop drinking or cut down in the past year — a measure, perhaps, of both self-recrimination and the "dryness" of drinking norms. Could it be that the more permissive normative structure observed for Scottish men explains this curious difference in symptom frequencies. But if so, why do Scottish females differ from U.S. females on this measure as much or more than Scottish males differed from U.S. males?

Weed Management

Lesson Introduction

Explain to students that a great deal is already known about noxious weed control. One of the aspects of control is that a multi-pronged attack is usually taken with chemical, biological, and cultural control methods all being used either simultaneously or in sequence. Much research centres around development of plant specific controls, cost of controls, and reducing "collateral damage" that is inherent in the control methods utilized. Be sure to explain that the cost of fuel is always a consideration, whether it is fuel to run the spray rig and tractor, fuel to deliver the goats or bugs to the site and haul away the weeds you pull from the ground. This section of lessons has the opportunity to be expanded greatly. The two CD-Rom disks are very detailed and each section of each disk gives students a chance to study a different aspect of weed management. If students acquire current values for products and services, they can gain a greater understanding of the impact that weeds have on ranching, especially on cattle production.

Montana Education Content Standards & Benchmarks

Vocabulary

Acre – *noun*, 43,200 square feet or 1/640 of a square mile (5280' X 5280').

AUM – *noun*, animal unit month, amount of grass one cow eats in one month (about 600 pounds).

Biological control – *noun*, the type of living organism used to control the number of weeds, such as flies, sheep, and fungi.

Chemical control – *noun*, a process of controlling the number of weeds by spraying chemical herbicides on the plants. These chemicals are usually specific for broadleaf plants.

Cultural control – *noun*, a process of controlling the number and spread of weeds through plowing, pulling, mowing, burning, or irrigating.

Entomologist – *noun*, a scientist who studies insects and their life cycles.

Fungus – *noun*, a plant that lacks chlorophyll and obtains its nutrients from a host organism.

Gall fly – *noun*, a biological control that reduces the number of seeds produced by knapweed plants. Gall flies lay eggs in knapweed seed heads.

Herbicide – *noun*, a chemical used to control, suppress, or kill weeds by severely interrupting their growth process.

Infestation – *noun*, a troublesome or hostile spreading that threatens to overrun or inhabit an area in such large numbers that other organisms could be harmed.

IPM – *noun*, Integrated Pest Management, a program for weed control that uses all the available methods of control, rather than using just a single method.

Irrigate – *verb*, supply water to an ecosystem or crop land artificially to increase the productivity of the land.

Larva – *noun*, the immature, wingless, and often worm-like feeding form that hatches from the egg of many insects.

Section – *noun*, 640 acres or one mile X one mile.

Student Activities

1. Read and understand the "Montana Agricultural Statistics" pdf files. Research current costs and values of items mentioned and insert those present values and costs in place of those given in the handout.
2. Read the "Joe Smith Ranch Information Sheet" and calculate the results of the problems shown in questions 1-7 from. Show and label all work.
3. Read the "Chemical Weed Control Information Sheet" and answer questions 1-6. Show and label all work.

Cultural Control Strategies

Mulching

Mulches of wood chips, straw, grass clippings, paper, rocks, rubber, black plastic, or landscape fabric can be used to control weeds. Mulching is typically done on relatively small areas where the application can be monitored. Mulching works best on annual weeds and cannot control some perennial weeds because their extensive food reserves allow them to continue to grow up through the mulch. Wood chips are often used in forestry/wildland settings where woody weeds and undesirable vegetation can be shredded into wood chips and left onsite as mulch. In order to be effective as a weed barrier, the mulch should be 3 – 6 inches thick. Landscape fabrics are also used as mulches but are more commonly used on a smaller scale as they tend to be more expensive. These fabrics are heavy, sometimes woven and are made of polypropylene or other weather-resistant materials. Water passes through them slowly and many are made to prevent weeds from growing through them. When using organic mulches like wood chips, straw, or grass make sure the material comes from a weed free source and/or does not contain any weed seeds or fragments.

Burning

The most effective fires for controlling invasive plant species are typically those administered just before flower or seed set, or at the young seedling/sapling stage. Sometimes prescribed burns that were not originally designed to suppress an invasive species have that happy side effect. But in some cases, prescribed burns can unexpectedly promote an invasive, such as when their seeds are specially adapted to fire, or when they resprout vigorously. These prescriptions must be modified or other management actions taken to undo or reverse the promotion of the invader.

Most successful weed control efforts that result from burning are due to the restoration of historical (natural) fire regimes, which had been disrupted by land use changes, urban development, fire breaks, or fire suppression practices. Many prescribed burn programs are, in fact, designed to reduce the abundance of certain native woody species that spread into unburned pinelands, savannas, bogs, prairies, and other grasslands. Repeated burns are sometimes necessary to effectively control weedy plants, and herbicide treatments may be required to kill the flush of seedlings that germinate following a burn.

Spot-burning invasive weeds with a propane torch can be cheaper and easier than implementing a prescribed fire (permits are still required), but is only effective when the infestation is small. Spot-burning can be used to burn individual plants, groups of plants in a small area, or to ignite brush piles. Propane torches can be used in areas where there is little or no fine fuel to carry a prescribed burn, and can also be used to kill plants when conditions are wet.

Prescribed Burning and Herbicides

Some invasive species have underground storage organs that resprout vigorously after fire, and/or seeds whose germination is stimulated by fire. Some of these species may not be possible to control with fire, but some can be controlled with repeated burns and others may be especially vulnerable to herbicides after a burn. Resprouts or seedlings that are 1 to 3 months old are often especially sensitive to herbicides.

Timing of Burn

The timing of a burn can strongly affect the fire's impact on native and exotic plant populations. For example, in California's Carrizo Plain Natural Area, Meyer & Schiffman (1999) determined that warm-season prescribed burning (late-spring and fall) was most effective for reducing abundance of Mediterranean annual grasses. Native plant cover and diversity also increased significantly following warm-season prescribed burns. Winter burns, however, did not affect the abundance of native plants, and exotic plant cover was only moderately reduced.

Timing was also key in controlling smooth brome *(Bromus inermis)* and encouraging the growth of native grasses in Nebraska and Minnesota (Willson & Stubbendieck 2000). Timing prescribed burns so that they occurred at the time of tiller (aboveground lateral stem) elongation, yielded an immediate and persistent reduction in both tiller density and biomass of smooth brome.

Burning in Extensively Disturbed Areas

Not all burn treatments in wildlands are beneficial. When fires become too intense, crown-fires and death of native plants that typically survive fires can result. If temperatures are too hot, soil organisms and seeds, even those of species that require fire stratification for germination, may perish, and valuable soil nutrients may be volatilized or otherwise lost. In extensively disturbed areas of southwest Australia, fire actually enhanced the invasion of weeds along roadsides, and resulted in an overall decrease in the abundance of native species (Milberg & Lamont 1995). Schwartz & Heim (1996) reported that fire was at best moderately successful for garlic mustard (*Alliaria petiolata*) control in Illinois forests, and Luken & Shea (2000) determined that repeated prescribed burning had no significant effect on garlic mustard in Kentucky. In both cases, however, the burns were detrimental to native herbaceous species, reducing both density and richness. Even three years after the initial burns, native plant composition did not recover to pre-burn values.

Preventing Spread of Weeds

Keep all equipment, trucks, and engines clean of weed seeds. After each burn, and before moving to another site, be sure to clean (hose-off) all equipment, tools, and clothing used. This will minimize changes of carrying weed seeds directly to a new site where a fire might provide perfect conditions for their establishment.

Cultural Control of Established Dandelion in Turf

Dandelion can be a major weed problem in turf. It forms dense clumps which affects the footing and the overall playability of sports fields and golf courses. The texture and colour of its leaves does not blend harmoniously with the surrounding grasses. When in bloom, its yellow flowers are yet a further detractor from the aesthetics of the turf. Dandelion seed is spread on the wind, and as this seed can be carried great distances, prevention of new infestations is difficult. Once a dandelion becomes established and supported by its large taproot, they are even harder to control.

The foundation of cultural dandelion control in turf has always been to create a growing environment in which the turf has the competitive vigor to repel the establishment of new dandelion seedlings. However, when it comes to dealing with established dandelion, many of the cultural methods are abandoned, opting to solely rely on chemical control. As concerns over the use of herbicides for aesthetic purposes increases, the turf manager must explore new strategies and rethink the whole dandelion control process. In a previous study conducted by the Prairie Turfgrass Research Centre, mowing various species of grasses at a higher height was thought to reduce weed populations. A number of grasses resisted the development of weeds when mowed at 3" and there was virtually no weed development over the duration of the study. Aeration and topdressing are cultural practices that improve gas diffusion in soil which in turn helps to promote root growth. In addition, aeration physically removes thatch from the turf which also helps to improve the turf quality.

High quality turf as promoted by adequate fertility has also shown to reduce weed infestations. In addition, the two organic fertilizers, soybean and corn gluten meal, have shown to have weed suppression properties. This study was developed to determine the effects of various cultural strategies for the control of dandelion in turf.

Methodology

The effect of three cultural turf management practices on an established dandelion population was tested at the Prairie Turfgrass Research Centre on a site located at Olds College, Olds, Alberta. Two weeks prior to the establishment of the trial (July 8), a pretrial treatment of the herbicide Trillion was applied to the site in an attempt to suppress the dandelion infestation. Seven days after the herbicide had been applied the level of control was visually assessed. On June 22 nd the trial was established over the pretreated area and the location of each plot was identified and staked to allow for future reference. Test plots that measured 1x 1 metres were laid out on a turf site with a uniform stand of established dandelion. Each treatment was replicated four times within a three factor randomized complete block design.

To determine if soil aerification was a factor to consider when controlling established dandelions, half of the plots were aerified, while the other half was left untreated. A Ryan GA30 aerator, equipped with 5/8"hollow tines and set for a 3"x 3" spacing, was used to effectively open up the turf. The soil cores were harvested and removed from the site. USGA specification sand was spread over the aerified turf. Several passes with a drag mat were made to fill the core holes and to evenly distribute the sand over the aerated area. The second factor evaluated the effect of fertility on weed control. The individual fertilizers were weighed into plot sized lots and applied by hand using a simple shaker bottle. The plots received the initial application of fertilizer on the establishment date and a second application of the fertilizer was applied four weeks later. The mowing treatments were conducted on a weekly basis. This process was accomplished by first mowing the entire experimental area at the higher cutting height of 2½" (6.25cm), then using a push mower, selectively recut specific plots at the lower height of ¾" (1.9cm).

On August 31 st the trial was evaluated for the first time. The dandelions in each of the one square metre plots were physically counted. In addition the plots were visually evaluated for overall turf quality. Following National Turfgrass Evaluation Program (NTEP) protocols, three turf quality factors: colour, density and area cover were assessed. The colour factor subjectively evaluated the uniformity and intensity of the colour displayed by the turf. A 1 to 9 scale was used to rate each plot. Treatments which stimulated a uniform dark green colour received scores ranging from 6 for an acceptable colour to 9 for turf with outstanding colour. Plots that produced a lighter turf colour were scored lower. Density, the second quality factor, was subjectively evaluated for the impact of the treatments on the turf to produce more shoots and tillers. The 1 to 9 scale was again used to rate each plot. Treatments which stimulated a tightly knit turf received scores ranging from 6 for an acceptable density to 9 for superior turf. Treatments associated with a weak or thin turf stand were scored lower.

The final quality factor, area cover, subjectively evaluated the vigor of turf. Once again a 1 to 9 scale was used to rate each plot. Treatments which stimulated a thick competitive turf cover received scores ranging from 6 for an acceptable area cover to 9 for a superior area cover. Treatments producing a weak turf, affected by weed encroachment and/or the presence of bare patches, were scored lower.

Results

Herbicide Pretreatment

Seven days after the herbicide had been applied the level of control was visually assessed.

Classic signs of herbicide damage were evident throughout the trial site as the top growth of the dandelions appeared twisted and some upper leaves were beginning to turn brown. At this point, good control was expected. However, as dandelions counts ranged from 2229 dandelions per m 2 on the August 29 rating date, effective control was questioned.

Mowing Height

An analysis of the data indicates that there was no statistical difference in the dandelion population when the two mowing heights were compared. Even though the turf mowed at the higher height scored better than the turf mowed at the lower height for colour, density and area cover, it was not statistically different than the turf that was mowed at a lower height.

Turf Aeration

At the ninety percent confidence level the turf which was cored and top dressed had a significantly higher number of dandelions per square metre than the nonaerated turf. The turf colour and turf density of the nonaerated turf was significantly better from that of the aerated turf. There were no significant differences between the treatments for area cover.

Fertilizer

At the ninety percent confidence level the high rate of the controlled release synthetic fertilizer had fewer dandelions per square metre than the other fertilizer treatments. The turf colour stimulated by the high rates of the corn gluten and the controlled release synthetic was significantly better than the turf colour of the unfertilized treatment and the lower rates of either the soybean meal or the controlled release synthetic fertilizer.

Discussion

In the initial year of this study, the infestation of dandelion had remained very high. The pretrial herbicide treatment provided only short term suppression of the dandelion's top growth. With only a single set of data points it is still unclear as to which treatments or combination of treatments will have the greatest effect on the dandelion population.

Cultural Control of Herbicide-resistant Lolium Rigidum Gaud. Populations in Winter Cereal in Northeastern Spain

Lolium rigidum Gaud. is one of the most common weed species in winter cereals in Northeastern Spain. Herbicide resistance of *Lolium rigidum* Gaud. to chlortoluron and diclofop-methyl has been reported in Catalonia since 1995 (Taberner *et al.*, 1995). The problem has continued growing and more populations have been found resistant to tralkoxydim, fluazifop and chlorsulfuron (Taberner, 2005). The estimated affected area is still less than 0.5% of the 6 million hectares of winter cereal in Spain (Díaz and Gorrochategui, 2005) but the affected fields require costly management. Low cereal yields and selling prices have encouraged Spanish farmers to invest minimum efforts in their cereal crops spraying expensive graminicides only when the cheap ones failed. However, the current high cereal price situation may introduce new elements in this scenario modifying the prospects for herbicide resistance.

The great capacity of *L. rigidum* for developing resistance adapting to many different compounds has led to the development of multiple resistant populations in Australia (Powles, 1993) and to the existence of biotypes resistant to herbicides belonging to 10 different mode of action groups (Heap, 2005). Taking into account that climatic and edaphic conditions of some regions in Australia and Spain are similar, the herbicide resistance problem can potentially grow in Spain. In order to prevent and control this problem it is necessary to develop non-chemical control strategies, adapted to the local farmers' practices (Matthews, 1996; Gill and Holmes, 1997; Nalewaja, 1999). Out of these options, probably the easiest to be adopted in the Spanish context are delayed sowing, increasing weed density and ploughing. Stubble burning and crop topping are not authorised, pastures are not sown, and the resistance problem is probably not serious enough to justify modifying the external combine harvesters for seed catching as proposed by Matthews *et al.* (2004). If the problem grows, as it occurred in Australia, Spanish farmers will probably also adopt more non-chemical management strategies, improving these methods even years after the outcome of the problem, as exemplified with the seed burner presented by Matthews *et al.* (2004), thus showing that there is still interest in offering new alternatives to be combined with herbicides. Two main aspects of the seed biology of *L. rigidum* offer opportunities for its control. First, it is known that the viability of the seeds in the soil is short (Gramshaw and Stern, 1977; Taberner *et al.*, 1992) so that burying them by ploughing can reduce their viability rapidly. For seeds buried at 0, 5 or 20 cm depth, Taberner *et al.* (1992) found that one year afterwards only around 5% of the seeds remained viable in Spanish conditions. This behaviour opens the possibility of burying the seeds with a deep ploughing.

A second important characteristic of *L. rigidum* is that main germination occurs after rainfall in the beginning of the season (Gramshaw and Stern, 1977). Cheam *et al.* (1998) observed main emergence in the two weeks after crop sowing and up to 93% emergence concentrated in the 5-6 first weeks, while Matthews (1996) (using data of Heap, 1988) describes how 60-80% of the *L. rigidum* seeds emerged in the initial germination flush in Australian conditions.

This fact offers the opportunity of delaying the sowing date controlling the first *L. rigidum* emergence flush before sowing. However, depending on the climatic conditions, germination can also be staggered. In some cases, in Spanish conditions, there is a concentrated germination flush during the first 40 days after cereal sowing (Planes *et al.*, 1999) but in other cases the germination period can start 2-4 weeks after sowing and last 4-7 weeks (Recasens and Kuc, 1998). Sowing delay aiming a reduction in weed populations can thus be very variable, depending on the climatic conditions of that year. An important inconvenient of sowing delay is the possible yield decrease (Planes *et al.*, 1999; Anderson, 2007).

Lolium rigidum control caused by increased sowing density has been scarcely reported. Gill and Holmes (1997) stated that Australian farmers are usually not willing to increase wheat plant density above 100-150 plants m-2.

In Central Spain, Lacasta *et al.* (2004) tested the effect of sowing barley at three different densities (80, 160 and 240 kg ha-1) during eleven years in a monocrop situation finding that the intermediate rate (160 kg ha-1) generally yielded the most. In irrigated winter cereal, Moreno *et al.* (2002) recommended using more than 100 kg ha-1 in barley, as they did not find any significant yield increase in any of the three years after testing five different rates up to 200 kg ha-1 in the semiarid Spain. In different experiments at variable *L. rigidum* density, Izquierdo *et al.* (2003) did not find significant differences in barley yield for densities between 75 and 300 kg ha-1.

Liebman and Gallandt (1997) recommended combining several control techniques to have additive 148 *A. Cirujeda and A. Taberner/* Span J Agric Res (2009) 7(1), 146-154 effects, the so called 'many little hammers'. Similarly, Anderson (2007) proposed the dualistic approach of prevention and control, which is apparently effective in the semiarid steppe of the United States. However, little work on these techniques have been conducted in our area.

The objectives of this work were to test the effect of sowing delay, ploughing, increased cereal rate and the combination of these methods on *L. rigidum* control in winter cereal. The final aim was to provide new data on the effects of these methods, which should be combined in an Integrated Weed Management Programme to control herbicide resistant *L. rigidum*.

Material and Methods

Plant Material

Field trials were conducted on different herbicideresistant populations at different locations of the Lleida province. All experiments were conducted on commercial fields using the varieties chosen by the farmer and applying conventional fertilization. The resistance to herbicides was checked before treatment using a seed-based quick-test and confirmed in herbicide field trials. All the populations were resistant to chlortoluron, diclofop-methyl and tralkoxydim at the same time, excepting the population of barley in Sarroca, which was only resistant to chlortoluron. *Lolium rigidum* plant density in the control plots was high in all cases. Plots were placed perpendicular to the straw swaths, where this weed emerges predominantly (Blanco-Moreno *et al.*, 2004) so that *L. rigidum* distribution was similar between plots.

Cultural Control Methods

Sowing delay was conducted at five fields; mouldboard ploughing at four fields; the combination of both practices at two fields; increasing the cereal density and increasing the cereal density combined with sowing delay at one field. The experimental design was a split-plot with ploughing type as a main factor at Ferran 1, Verdú and Concabella. At Santa Maria and Ferran 2, the treatments were arranged as a randomised complete block design with two factors. At Sarroca, the experiment was a randomised complete block design with one factor. Elementary plots measured 4 x 5 m in all cases. Ploughing was conducted with a five-furrowmouldboard plough at Ferran 1, Concabella and Verdú at 15, 20 and 25 cm depths, respectively. At Ferran 2 a four-furrow plough was used inverting the soil approximately at 18 cm depth. Commercial ploughs were used in all cases. At Santa Maria, normal sowing density was 175 kg ha-1 corresponding to 439 seeds m-2 and high sowing density was 1.5 fold the normal rate, that is, 250 kg ha-1 corresponding to 627 seeds m-2. Normal densities in the area were used in the rest of locations.

Assessments

Lolium rigidum density was assessed at weed and crop heading (BBCH scale 56-59, BBCH Working Group, 1997) to describe the final effect on the weed population. Assessments included number of *L. rigidum* plants, number of *L. rigidum* ears and total fresh biomass in 0.1 m2 squares. Biomass was determined immediately after cutting in the field. Crop plant density per linear meter was assessed three times in each plot at heading at Santa Maria (28th May 2003), Verdú (17th June 2003) and at Sarroca (27th April 2004) to describe the effect of the cultural methods on the crop.

Data Analysis

Data were transformed when necessary to satisfy normality and variance homogeneity following the indications of the Box-Cox Transformation (Bowley, 1999) using $(x + 1)0.5$ for most cases, excepting weed weight per plant at Ferran 1 where $x0.5$ was used. Analyses of variance were performed using the SAS GLM procedure (SAS Institute, 1991) for each location separately considering the different models. Least significant differences (LSD) were calculated for pairwise comparisons. Mean values are shown as least square means (LSM) and differences between means were analysed on the basis of ordinary *t*-tests considering $P < 0.05$ in the case of significant interactions.

Results

Mouldboard Ploughing

Mouldboard ploughing reduced significantly *L. rigidum* density at Concabella and Verdú. This difference was particularly remarkable at Concabella, where ploughing reduced 96% weed density. Although ploughing reduced 55% weed density at Ferran 1, this difference was not significant. Similar results were obtained for weed biomass in these three sites. No significant differences in densities and biomass were observed at Ferran 2. The lack of efficacy of ploughing at this site is probably due to the high stoniness at this location, which impeded deep enough ploughing without damaging the tool. Ploughing did not affect wheat density at Verdú, the only site where this parameter was measured.

Sowing Density

Increasing sowing density had no effect on crop density and on *L. rigidum* density and biomass at Santa Maria, the only site where this comparison was conducted.

The Combination of Different Techniques

The combination of sowing delay with ploughing or with sowing density did not improve the effect of the individual treatments significantly at any of the three sites but had some effect at Ferran 1, where *L. rigidum* density was reduced. At Ferran 1 and Verdú, ploughing was the most important control method. At Santa Maria, the main weed reduction was due to sowing delay.

Concerning the effect of these techniques on the crop, at Santa Maria, increasing the sowing rate at the second sowing date gave significantly higher wheat density than the other treatments. However, no effect was observed at Verdú on the crop due to the combination of both techniques.

Discussion

Efficacy on weed plants due to ploughing was not considerable at Ferran 2, around 50% at Ferran 1, 81% at Concabella and 93% at Verdú. These values are similar to those obtained by Recasens *et al.* (2001b) (90 and 81%) and by (Recasens *et al.,* 2003) (65 and 92%) in Northeastern Spain, a bit higher to those obtained by Gill and Holmes (1997) (13 and 64%) and a bit lower to those found by Matthews *et al.* (1996) (98%), in Australia. The variation between the experiments suggests that ploughing needs a precise implementation. It should be mentioned that a correct ploughing can be difficult in stony soils like the one in Ferran 2. High fuel consumption justifies its use only in extremely highinfested fields or if cereal price is high enough to compensate the ploughing costs.

Trial results confirm that climatic conditions in the time gap between the normal and the delayed sowing date are crucial for efficacy and should allow maximal *L. rigidum* emergence. Efficacy on weed plants was not considerable at Concabella and Verdú, 35% at Santa Maria, 54% at Ferran 1, and 88% at Sarroca. The highest efficacy was achieved in the field with the highest

Sowing Delay

Sowing delay affected significantly *L. rigidum* density at Ferran 1, Santa Maria and Sarroca. At Ferran 1 and Sarroca, rainfall between both sowing dates allowed new weed emergences, which were mechanically controlled, prior to the second sowing date. At Santa Maria, although little rainfall fell between both sowing dates the soil remained moist due to the abundant rainfall in the months before sowing and to the foggy days, allowing weed emergence. Total weed biomass was also significantly lower in the second sowing date at these three locations. In the other two trials located at Verdú and Concabella, there were no differences in either plant weed density and weed biomass between the two sowing dates. At these two locations too little rainfall fell before the first sowing date and between the two sowing dates to allow *L. rigidum* emergence. The overall efficacy of sowing delay in these trials was lower than the control obtained by ploughing. The effect of sowing delay on crop density was not statistically significant at Verdú, but at Santa Maria and at Sarroca significantly more crop plants established at the second sowing date ($P < 0.05$). At Santa Maria, no significant differences and even no clear tendencies on *L. rigidum* were detected due to increased sowing rate, so that probably wheat plants were not more competitive when sown in 1.5-fold the normal density.

A possible explanation is that the field at Santa Maria was irrigated, consequently, water was not a limiting factor. Increasing sowing density could have a more important effect on weeds in rainfed environments as found by Izquierdo *et al.* (2003) in a barley crop infested with *L. rigidum* who describe a lower weed biomass for the same number of weed plants. Taking into account the yield results of Moreno *et al.* (2002) and Lacasta *et al.* (2004), probably it is not worth to increase the sowing rate more than 160 kg ha-1 in these semi-arid environments.

The Combination of Different Techniques

The theory of using 'many little hammers' to suppress weeds (Liebman and Gallandt, 1997) was partly confirmed at Ferran 1 and also by Recasens *et al.* (2001a). However, the experience of Anderson (2007) in the semiarid steppe of the United States shows that more than two cultural control tactics may be necessary to be effective enough. In the semi-arid conditions of Spain, lack of water in the drylands hinders to integrate other tools proposed by Anderson (2007) as crop rotations of cold-and warm-season crops. In line with the experiences in his situation, no-till could be also used as an effective tool under Spanish conditions. Because seeds remain on the soil surface, they are vulnerable to specific mortality factors that act on surface lying seeds. Seed mortality could contribute declinining the weed seedbank. For example, Baraibar *et al.* (2008) found that predation by harvester ants (*Messor barbarus*) of *L. multiflorum* seeds was significantly higher in no-till than in conventional tilled fields. In case of combining sowing delay with herbicide use, an additional benefit of delayed sowing is that later emerged *L. rigidum* plants are weaker and probably more susceptible to herbicides or to other control strategies (Recasens *et al.*, 2001a).

The starting situation of the fields in the trials described in this work is of highly-infested fields with resistant weeds, which could not be controlled with the conventional herbicide use. In this context, weed density after combining two cultural control methods was still very high ranging between 35 and 450 *L. rigidum* plants m-2 depending on the field. Thus, in highlyinfested fields with herbicide resistant *L. rigidum,* it will probably be necessary to combine these cultural methods with other more drastic tools as leaving the field fallow for one year, depleting the seed bank by repeating cultivation as suggested by Gramshaw and Stern (1977), or combining them with the use of those herbicides that are still active on the populations. As commented previously, the new high cereal prices will probably favour the use of more expensive active ingredients, which should be used correctly in combination with other techniques without relying exclusively on them.

Taking into account that delayed sowing date can have negative consequences on yield due to lack of rainfall and that the effect on *L. rigidum* is more variable, mouldboard ploughing can be a more secure strategy for its control than sowing delay, with more constant effects on weed plant reduction and with a lower impact on yield. The main problem of this technique is the high fuel consumption, which needs to be compensated by the cereal price and by the *L. rigidum* control. Taking into account the fast decline of the seed bank of this species, it is probably useful to plough every 2-3 years, only, to reduce the resistant populations. As the highest reductions were achieved by combining ploughing with sowing delay, it is probably worth to practice both, though high densities of *L. rigidum* will require additional control methods.

Development of Cultural Control Methods for Pests of Leafy Vegetables

This project aimed to evaluate whether or not certain plantings of non-crop plants can assist with pest control in leafy-vegetables such as lettuce and celery. It also assessed the potential of other management practices such as addition of composted fowl manure and using flowering weeds to augment populations of predatory and parasitoids insects and mites. The target pests were primarily aphids and thrips. These are important groups for leafy vegetables as they include lettuce aphid (*N asonovia ribis-nigri*) and western flower thrips (WFT) (*Frankliniella occidentalis*), both of which are resistant to a range of insecticides.

WFT is also an important vector of tomato spotted wilt virus. Strip plantings of cereals will attract certain species of aphids which may be pests in some crops (such as wheat or barley) but are not pests in leafy vegetables. However, predators of cereal aphids such as brown lacewings (*Micromus tasmaniae*) are predators of pest aphids and other insect pests in leafy vegetables. We aimed to find out whether or not strip plantings of cereals within horticultural crops can (i) increase the populations of beneficial species and (ii) increase the level of control of pests. Mites, staphylinid beetles and predatory thrips are known to prey on thrips and this project investigated how populations of these groups could be encouraged. It has been noted recently (Beaulieu and Weeks 2007) that there are many soil-dwelling predatory mites that have potential as biological control agents in agriculture, but there has been extremely little research on this group. The use of composted fowl manure was seen as one potential means of achieving increases in resident, soil-dwelling predators. The sequence of events is hypothesised as follows:

1. The application of composted fowl manure will result in more sciarid flies.
2. Predatory mites and beetles that prey on these flies (as eggs and larvae) will be advantaged
3. The predatory mites that are encouraged feed on WFT
4. Therefore, improved control of WFT can be achieved by enhancing generalist predators including predatory mites.

There have been many studies on the provision of pollen and nectar sources to augment populations of biological control agents, particularly hoverflies. There are also some studies where "insectary plantings" have been used to provide an in-field resource to maintain populations of beneficial insects (Colley and Luna 2000). Examples of this type of approach from Australia include the use of strip plantings of lucerne in cotton (Cotton CRC 2005), grasses in citrus (Smith *et al.* 1997) and buckwheat in vineyards (Bernard *et al.* 2006, Berndt *et al.* 2002). In this project we conducted preliminary observations on the potential for existing weeds, particularly flowering aniseed weeds, to provide the same resource for beneficials.

Results

Insectary Plantings and Beneficial Species

The first significant result was that farmers were easily able to understand the concept of insectary plantings and were able to create them using ryegrass. Some farmers planted the entire paddock to cereals and then ploughed them in, others planted all sprinkler rows, and others planted strips throughout the crop. The decision on what sized strip to plant was made by the farmers on the basis of what was practical for them to manage. In all cases we were impressed with the fact that the growers were very happy to plant the cereals once they understood the logic behind the trials. It must be remembered that areas planted to cereals would not produce any crop for them. Sampling of lettuce was undertaken at 4 IPM sites and 6 non-IPM sites. WFT was not abundant in IPM grown lettuce but was found at a much higher level, causing visible feeding damage, on non-IPM lettuce. Two IPM sites were in Werribee South and two were in the Cranbourne area. All of the non-IPM sites were in Werribee South. It ranks the maximum counts of WFT in 10 harvested lettuce during the season from each site as either low (0-25 WFT total), medium (25 – 50 WFT total) or high (50 – 500 WFT).

Discussion and Conclusions

It is clear that there are several useful cultural control measures that are practical, that vegetable growers can use to encourage beneficial insects and mites. That is, the cultural and biological components of an IPM strategy can be strengthened. This project has demonstrated that it is possible to use insectary plantings and made a part of normal vegetable production methods. The use of composted organic material was also shown to increase populations of soil-dwelling predatory insects and mites. The beneficial insects and mites that are encouraged by these different methods fall into two categories, transient and resident. The strip plantings of cereals attract transient brown lacewings (*M. tasmaniae*) as they are dispersing through vegetable production areas in spring. They will stop in these strips but not in a bare paddock, and so it is a means to increase the population of predators on a paddock going into vegetable production. The organic composts can be used to increase the populations of resident soil dwelling predators. The cultural control component of IPM is often given less emphasis than the biological or chemical components but this project has shown that it provides the basis for control of key pests. There are many different types of cultural controls that are available to farmers that are beyond the scope of this project but many are site or crop-specific. This work has focussed on methods to increase populations of beneficial invertebrates. Providing a suitable habitat for predatory species increases the role of both the resident and transient biological control agents. Planting insectary crops has also the added benefit of helping make growers aware of the broader approach to controlling pests by deliberately acting to encourage beneficial species. It is therefore a good starting point for those just adopting IPM, to emphasise the importance of cultural controls within an IPM strategy.

There is also potential for certain existing flowering weeds to be used to increase populations of beneficial insects by providing pollen and nectar. The situation that exists in areas such as Werribee South is that on land adjacent to crops that is usually not owned or managed by the farmer, there are often flowering weeds of different species. Management of these is restricted to limited herbicide applications which can result in bare ground or poor control of established weeds. Replanting these strips of land with native plants may be desirable if particular species can be recommended, but no local species have been trialled in this area and it is currently not achievable by the grower, and is not a priority for them. This especially relevant for farmers leasing land on a short-term basis. We suggest here that there is scope to utilize some existing flowering weeds in preference to attempting to achieve bare earth. The main concern is whether or not certain species of weeds will also encourage pests or be a reservoir of disease. Therefore, in brassica production areas brassica weeds would not be suitable but aniseed weed could be ideal. However, in celery production areas aniseed weed could harbour celery mosaic virus and so would not be suitable. The observations in this project have been that there are often large strips of flowering weeds on land adjacent to crops anyway, but that some of them could be considered useful rather than just weeds.

Chapter

8

Pest Control Tactics

The history of pest control probably began with the first human who ever swatted a mosquito or picked off a louse. From the fossil record, we know that all major taxa of biting flies and external parasites already existed by the time *Homo sapiens* first appeared on earth. *Phthirus* and *Pediculus*, the two genera of lice that feed on humans, have a host range that is limited to primates (apes and monkeys). And we suspect that human fleas (*Pulex irritans*) and bed bugs (*Cimex lectularius*) adopted cave-dwellers as hosts because these insects are most closely related to other species that live on bats. But since our primitive ancestors were hunters and gatherers, they probably found that insects were more useful as food than they were troublesome as pests. (Even today, people in some primitive cultures eat the lice they pick from one another's hair). It was probably not until the dawn of organized agriculture, when insects attacked the plants we grew for food, that we first recognized them as a potential threat to our own survival.

Pest control tactics were mentioned occasionally in writings of the ancient Chinese, Sumerian, and Egyptian scholars. Many of these tactics were embedded in religion or superstition, but a few had real scientific merit. Predatory ants, for example, were used in China as early as 1200 B.C. to protect citrus groves from caterpillars and wood boring beetles. Ropes or bamboo sticks tied between adjacent branches helped the ants move easily from place to place. A passage in Homer's Iliad (8th century B.C.) describes the use of fire to drive locusts into the sea, and the ancient Egyptians organized long lines of human drovers to repel swarms of invading locusts. Pythagorus, a Greek philosopher and mathematician, was credited with clearing malaria from a Sicilian town during the 6th century B.C. by instructing its residents to drain the marshes. Chemical substances that purportedly killed or repelled insects were in common usage. Many of these were of questionable value, but some worked, and a few are still in use today. Some of the inorganic compounds, such as sulfur and arsenic, have well-established insecticidal activity. And modern science has only recently come to recognize that many plant extracts used by ancient apothecaries (e.g, lemon oil, wormwood, hellebore, fleabane, etc.) do indeed contain compounds with useful activity against insects.

There was very little progress in pest control during the dark ages. Ignorance and superstition abounded. For what it was worth, St. Bernhard excommunicated the flies of his parish in 1121. In a book entitled "Natural History", Ferrante Imperato (1599) gave a prescription for eliminating flies from a dwelling:

> *"... draw the image of a fly... on a copper plate during the second half of the constellation of Pisces... then bury it in the centre of your house (during) the first half of the constellation of Taurus."*

With the Renaissance, people began to view insects less as a punishment from God and more as members of a natural world that could be studied and controlled. A renewed interest developed in insects both as organisms and as pests. More accurate observations of natural history and behaviour led to more inventive control practices. Hand labour was used extensively in early pest control, but cultural, physical, and chemical practices also evolved. Franz Ernst Brückmann, a German physician who lived in the early 1700's, designed the first mechanical traps for various insects. His fly traps consisted of a wooden box baited with a sweet attractant and equipped with a spring-loaded lid. When several flies had settled inside the trap, the lid was snapped shut to trap the flies inside. The volume of the trap was then decreased by sliding the ends of the box together and squashing the flies against their bait. Brückmann also designed flea traps. They were hollow, perforated cylinders, baited with blood or honey, and worn around the neck as a pendant. For a time, these flea traps, crafted from ivory or silver and often ornate in design, were popular fashion accessories worn by the aristocracy of western Europe.

Since the late 1800's, entomologists and chemists have made outstanding progress in the technology of pest control. Today's arsenal of weapons is large and diverse, encompassing legal, cultural, physical, genetic, and biological tactics, in addition to the well-known chemical pesticides. In general, all of these tactics work in at least one of the following ways:

- They kill the pest directly — usually by exposing it to lethal substances or unsuitable environmental conditions.
- They reduce the reproductive potential of a pest population — often by modifying its environment (biotic or abiotic) or by restricting its movement.
- They modify the pest's behaviour to make it less troublesome (attract it, repel it, confuse it, exclude it, mislead it).

Most of the control tactics that are commonly used today can be grouped into two broad categories: natural controls and artificial controls. By definition, a natural control may be any environmental factor that keeps a pest population below its economic injury level. Examples might include geographic barriers, cold temperatures, or natural enemies that keep population growth in check. Artificial controls, on the other hand, employ products or processes of human origin to modify a pest's distribution, behaviour or physiology. Fly swatters and insecticides are two obvious examples. But as is often true in biology, there are some control strategies that seem to straddle the line, exhibiting characteristics of both groups or not fitting neatly into either category. In fact, some of our most effective tactics are natural controls that we can improve, enhance, accelerate, or augment by appropriate human intervention.

Cultural Control

Surprisingly simple modifications of a pest's environment or habitat often prove to be effective methods of pest control. As a group, these tactics are usually known as cultural control practices because they frequently involve variations of standard horticultural, silvicultural, or animal husbandry practices.

Since these control tactics usually modify the relationships between a pest population and its natural environment, they are also known, less commonly, as ecological control methods. Simplicity and low cost are the primary advantages of cultural control tactics, and disadvantages are few as long as these tactics are compatible with a farmer's other management objectives (high yields, mechanization, etc.). Unfortunately, there are still a wide variety of insect pests that cannot be suppressed by cultural methods alone.

Crop rotation is one of the oldest and most effective cultural control strategies. Growing a single crop year after year in the same field gives pest populations sufficient time to become established and build up to damaging levels. Rotating the field to a different type of crop can break this cycle by starving pests that cannot adapt to a different host plant. Farmers in the Midwest, for example, can reduce populations of wireworms (Elateridae) and rootworms (*Diabrotica* spp.) in corn fields by switching to oats, wheat, or legumes. Similarly, the clover root curculio (*Sitona hispidula*) that feeds exclusively on legumes can be eliminated by switching from clover or alfalfa to corn or small grains. Rotation schemes have also proven successful for controlling pests in pasture lands. Texas cattle fever, a protozoan disease transmitted by cattle ticks, *Boophilus annulatus*, has been eliminated from the southern United States due, in part, to a diligent campaign of pasture rotation designed to protect beef and dairy herds from tick infestation.

Crop rotation schemes work because they increase the diversity of a pest's environment and create discontinuity in its food supply. As a rule, rotations are most likely to be practical and effective when they are used against pests that:

a. attack annual or biennial crops
b. have a relatively narrow host range
c. cannot move easily from one field to another, and
d. are present before the crop is planted.

Intercropping (also known as mixed cropping) is another way to reduce pest populations by increasing environmental diversity. In some cases, intercropping lowers the overall attractiveness of the environment, as when host and non-host plants are mixed together in a single planting. But in other cases, intercropping may concentrate the pest in a smaller, more manageable area so it can be controlled by some other tactic. Strips of alfalfa, for example, are sometimes interplanted with cotton as a trap crop for lygus bugs (Miridae). The alfalfa, which attracts lygus bugs more strongly than cotton, is usually treated with an insecticide to kill the bugs before they move into adjacent fields of cotton.

In some crops, it is possible to create discontinuity in the pest's food supply simply by altering the time of year for planting or harvesting. This strategy, often known as phenological asynchrony, allows farmers to manage their crop so it remains "out of phase" with pest populations. Sweet corn, for example, can escape most injury from corn earworms (*Helicoverpa zea*) if it is planted in early spring and harvested before larvae mature. In the Midwest, farmers delay planting winter wheat until after the "fly-free date" to protect their crop from injury by the Hessian fly (*Mayetiola destructor*). These planting dates are calculated to ensure that wheat is not mature enough to attract egg-laying flies, yet still has enough warm weather to grow sufficiently before winter. Careful timing of harvest dates in alfalfa can cause high mortality in populations of alfalfa weevil (*Hypera postica*) and alfalfa caterpillar (*Colias eurytheme*) by removing edible foliage before the larvae have completed development. Harvest timing is often the most practical method available to foresters for controlling bark beetle (Scolytidae) infestations in pine plantations. Since trees become more susceptible to beetle outbreaks as they grow older, good management dictates that timber stands be harvested for lumber before the trees reach full maturity.

Managed application of water or fertilizer can have a big impact on the survival of pest populations in some crops. Annual flooding, for example, is a cultural practice that eliminates many potential pests in cranberry bogs. Regular rainfall (or overhead irrigation) can significantly reduce infestations of twospotted spider mites (*Tetranychus urticae*) in tree fruits. Good irrigation and appropriate fertilization keeps plants healthy, vigorous, and more resistant to insect injury. It is not unusual for small amounts of injury to actually stimulate compensatory growth in healthy plants.

Sanitation is another cultural control strategy that may be highly effective for some pests. Removing crop debris from cotton fields after harvest eliminates overwintering populations of pink bollworms (*Pectinophora gossypiella*), European corn borers (*Ostrinia nubilalis*), and sugarcane borers (*Diatraea saccharalis*). Collecting dropped fruit from beneath an apple tree reduces the next season's population of apple maggots (*Rhagoletis pomonella*), codling moths (*Cydia pomonella*), and plum curculio (*Conotrachelus nenuphar*). Shredding or burning the pruning wood from a peach orchard kills shothole borers (*Scolytus rugulosus*) and lesser peachtree borers (*Synanthedon pictipes*) that would otherwise emerge and reinfest the orchard. Clean cultivation is often recommended as a way to eliminate shelter and/or overwintering sites for pest populations. Simply tilling or plowing a corn field before winter may disrupt a pest's life cycle by causing mechanical injury, by increasing exposure to lethal cold temperatures, by intensifying predation by birds or small mammals, or by burying the pests deep beneath the soil surface. Populations of corn earworms and European corn borers have been greatly reduced in recent years by community-wide efforts to plow under corn stubble after harvest.

Although it works well in corn and cotton, clean cultivation is not always the best solution for eliminating pest populations. In some cases, ground cover or crop debris shelters natural enemies that are important members of the agroecosystem. Without adequate shelter these beneficial populations would die or move away.

Biological Control

Natural control strategies that employ biological agents for pest suppression are generally classified as biological control tactics. In conventional usage, this term usually refers to the practice of rearing and releasing natural enemies: parasites, predators, or pathogens. A slightly broader definition of "biocontrol" includes any related management activity that is designed to protect or conserve natural enemies.

Biocontrol agents include a wide variety of life forms, including vertebrates, invertebrates, fungi, and microorganisms. These beneficial species are common in most natural communities and, although their presence is often unnoticed, they help maintain the "balance of nature" by regulating the density of their host or prey population. Insect species often become "pests" when this ecological balance is disrupted by natural events or human intervention. Biological pest control strives to reestablish this balance in one of three ways:

Importation. Foreign exploration is conducted to identify and collect natural enemies in the country from which an exotic pest has been introduced. Following the discovery of a potential biocontrol agent, it undergoes extensive evaluation to insure that its ecology and host range are compatible with the community to which it will be introduced and that it will not become a pest once it is released. Suitable candidates are reared and released in the new habitat in hopes that they will become established and suppress the pest population.

1. *Conservation.* A variety of management activities can be used to optimize the survival and/or effectiveness of natural enemies. Conservation activities might include reducing or eliminating insecticide applications to avoid killing natural enemies, staggering harvest dates in adjacent fields or rows to insure a constant supply of hosts (prey), or providing shelter, overwintering sites, or alternative food sources to improve survival of beneficial species.

2. *Augmentation.* Natural enemies that are unable to survive and/or persist in a new environment can sometimes be reared in large numbers and periodically released to suppress a pest population. In some cases, small numbers of a beneficial species are released in several critical locations to suppress local pest outbreaks (an inoculative release). In other cases, larger numbers are released in a single location to flood the pest population with natural enemies (an inundative release).

Examples of Biocontrol

Just because an insect is a predator or a parasite does not necessarily make it a good biological control agent. Since the early 1900's over 600 species of beneficial insects have been introduced into the United States. Of these, only about 20% have been outright successes as biological control agents; another 35% have been partially successful, and the other 45% never became established or failed to have any significant effect on pest populations. From these successes and failures, entomologists have concluded that introduced biocontrol agents are most effective when they exhibit the following characteristics:

Narrow host range. Generalized predators may be good natural enemies but they don't kill enough pests when other types of prey are also available.

1. *Climatic adaptability.* Natural enemies must be able to survive the extremes of temperature and humidity that they will encounter in the new habitat.
2. *Synchrony with host (prey) life cycle.* The predator or parasite should be present when the pest first emerges or appears.
3. *High reproductive potential.* Good biocontrol agents produce large numbers of offspring. Ideally, a parasite completes more than one generation during each generation of the pest.
4. *Efficient search ability.* In order to survive, effective natural enemies must be able to locate their host or prey even when it is scarce. In general, better search ability results in lower pest population densities.
5. *Short handling time.* Natural enemies that consume prey rapidly or lay eggs quickly have more time to locate and attack other members of the pest population. Small populations of efficient natural enemies may be more effective biocontrol agents than larger populations of less efficient species.

6. *Survival at low host (prey) density.* If a natural enemy is too efficient, it may eliminate its own food supply and then starve to death. The most effective biocontrol agents reduce a pest population below its economic threshold and then maintain it at this lower equilibrium level.

Just as predators and parasites may serve as effective biocontrol agents for insect pest populations, so insect herbivores have also proven effective as agents for controlling weed populations. Feeding injury by herbivores may weaken a plant or increase its susceptibility to disease, desiccation, freezing temperatures, or other herbivores. In any case, an injured plant is likely to be a poor competitor, and in time, beneficial species will have an opportunity to reclaim the vacated space.

Insect Weed-eaters

The Klamath weed (*Hypericum perforatum*) and the prickly pear cactus (*Opuntia* spp.) are two well-known examples of successful weed control using insect herbivores. Biological control is a particularly appealing pest control alternative because, unlike most other tactics, it does not always have to be reapplied each time a pest outbreak occurs. Once natural enemies are released into a new environment, there is a good chance they will become established and provide a self-perpetuating form of control. Biological control is also the only control tactic that increases, rather than decreases, the species diversity within an agroecosystem. This increased diversity often results in greater stability because wild fluctuations in population density are less common in communities with a diverse food web.

Biological control is not a "quick fix" for most pest problems. Natural enemies usually take longer to suppress a pest population than other forms of pest control and farmers often regard this as a disadvantage. It also may be difficult to "integrate" natural enemies into a crop or commodity when pesticides are still in use. Beneficial insects are often highly sensitive to pesticides and their resurgence (recovery to pre-spray densities) is usually much slower than that of pest populations. Rapid pest resurgence often leads to a vicious cycle of continued chemical usage that prevents natural enemies from ever becoming reestablished.

Perhaps the greatest potential for future progress in biological control lies in improving the success of microbial pathogens. Many of these organisms are highly desirable as biocontrol agents: they attack a narrow range of insect hosts, they are not hazardous to humans or domestic animals, and they do not pose a threat to the environment.

As a group, though, pathogens have never proven to be quite as reliable as other forms of pest control. They are vulnerable to desiccation, ultraviolet radiation, and high temperatures. They may not survive long enough in the environment to encounter a host, and even if they survive, their virulence may be too low to overcome the host's defenses.

In the past few years, however, there has been encouraging progress in the development of microbial insecticides. These are commercial suspensions of spores, toxins, or virus particles that can be mixed with water and sprayed onto crops just like conventional insecticides. In many cases, microbial insecticides are better than conventional insecticides because they suppress pest populations without eliminating natural populations of predators and parasites.

A great deal of research is still needed before we can begin to benefit from the full potential of biocontrol. Fortunately, new developments in biotechnology may soon enable us to create new strains of microbial pathogens that are more virulent, easier to mass produce, and less sensitive to temperature and humidity. Work is currently underway to transform a common soil microbe, *Pseudomonas fluorescens*, into a pathogen of soil-dwelling insects by implanting the delta-endotoxin gene from *Bacillus thuringiensis*. Clearly, there are major changes in biocontrol that lie just over the horizon.

Breeding for Host Resistance

Breeding plants (or animals) for resistance to insects is really just another form of biological pest control. Rather than finding insects to attack the pests, breeders look for genetic traits (or combinations of traits) that reduce an organism's susceptibility to attack or injury by its insect pests. This idea was first tested in the 1870's by C. V. Riley, an entomologist who successfully fought a French outbreak of phylloxera (an aphid-like pest of grapes) with resistant North American rootstocks. Indeed, much of the pioneering work in this field has been done with crop plants because they are usually cheaper to breed and easier to manage than livestock. Recent innovations in genetic engineering, however, are likely to accelerate development of resistant genotypes in poultry, swine, and other commercial livestock.

In general, there are three approaches that plant breeders use to develop resistant cultivars:

1. *Antibiosis*. Plants produce a wide variety of defensive compounds (allelochemicals) that protect them from herbivores.

These compounds may reduce growth, inhibit reproduction, alter physiology, delay maturation, or induce various physical or behavioral abnormalities in herbivores. By purposely selecting for plants with high levels of allelochemicals, or by breeding such plants with less resistant ones, it is often possible to develop new cultivars that resist pest injury yet still retain desirable horticultural characteristics.

2. *Antixenosis.* A physical or chemical property of a plant can make it so unpalatable that it is largely protected from herbivore attack. This type of resistance is often known as nonpreference. It may involve the presence of feeding repellents (or the absence of feeding attractants), or it may involve physical traits such as hairs, waxes, or a thick, tough epidermis that do not provide the pest with a desirable feeding substrate. Alfalfa, for example, has been bred with hairy leaves to deter feeding by the spotted alfalfa aphid.
3. *Tolerance.* Some plant genotypes are simply able to "tolerate" injurious insects better than others. Tolerant cultivars may be exposed to the same pest populations as susceptible ones, but they do not suffer as much injury. Many varieties of field corn, for example, have fairly narrow, brittle stalks. When attacked by European corn borers, these stalks are further weakened and break easily in a windstorm. This wind damage, known as lodging, makes the corn hard to pick with mechanical harvesters. Plant breeders have largely resolved this problem by breeding for corn with thicker, stronger cornstalks. These tolerant cultivars are still attacked by corn borers, but they "stand up" to the injury and insure a harvestable crop.

Developing resistant cultivars by traditional breeding methods can be a slow and uncertain process. But new advances in biotechnology now make it possible to pluck genes from one organism and insert them into the cells of a completely different species. For example, the delta-endotoxin gene from *Bacillus thuringiensis*, an insect pathogen, has already been isolated and inserted into crop plants (notably cotton and tobacco). Cells of the genetically engineered plants can produce their own toxin, making them "resistant" to certain insect herbivores (e.g., Lepidoptera). Although gene-splicing techniques are still in their infancy, they offer exciting new possibilities for expanding the diversity of resistant plant genotypes.

Legal and Regulatory Control

At first, it may seem ludicrous to even contemplate the use of legal measures as pest control strategies — it's easy enough to pass laws, but how do you train the insects to obey them? In reality, legal sanctions are aimed, not at the insects themselves, but at a range of human behaviors that are most likely to affect the dynamics of pest populations. Legal control tactics, therefore, include all forms of legislation and regulation that might prevent establishment or reduce spread of an insect population.

The Plant Quarantine Act of 1912 was the first legal action taken in the United States to prevent the introduction of pests from foreign countries. This law, and others that followed it, established a network of inspection stations at major ports of entry and gave the federal government authority to organize border quarantines, to inspect all agricultural products, and to restrict entry of any infested goods. Today, these inspections stations operate under the jurisdiction of the Animal and Plant Health Inspection Service (APHIS), a branch of the U. S. Department of Agriculture. APHIS operates about 85 inspection facilities, employs 400-500 inspectors, and intercepts over 20,000 potential pest introductions in an average year. It also operates the Plant Quarantine Training Centre where inspectors from all over the world are trained to spot problems before they cross international borders. These agents, now representing nearly 40 countries, cooperate in efforts to limit the spread of indigenous pests under an international agreement that requires inspection and certification of nearly all agricultural commodities prior to export.

Federal import/export laws have been enacted to establish standards for a wide range of agricultural products. Most of the fruits and vegetables imported to the U.S., for example, must receive some form of fumigation, heat treatment, controlled atmosphere storage, or radiation exposure before they are released from quarantine. These post-harvest treatments are designed to kill whatever pest species may be present without affecting taste or quality of the produce.

Despite the best efforts of APHIS inspectors, some pests still slip into the U. S. and become established (e.g., cereal leaf beetle), or move naturally across international borders (e.g., imported fire ants and Africanized honey bees). When this happens, APHIS cooperates with state and local governments to establish domestic quarantines for limiting the movement of these pests within our borders. Whenever appropriate, eradication programs may be launched together with quarantine actions in an effort to destroy the entire pest population while it is still small. Inspections and quarantines are also operated by some state and local government agencies. The California Border Patrol, for example, is notorious for its interceptions of uncertified agricultural products at the state line. The rigid laws that prohibit private citizens from bringing a bag of oranges into southern California are generally given credit for excluding numerous pest species that might have caused millions of dollars in losses.

The price tag for these prevention efforts is only a fraction of the estimated $350 million Californians spent to clean up their most recent infestation of the Mediterranean fruit fly (*Ceratitis capitata*). Since 1915, nearly 300 pest species have been under some sort of domestic quarantine. The success of these programs has been erratic. In some cases (e.g., screwworm flies, khapra beetles, and parletoria date scale), eradication programs have managed to eliminate the pests. But for every noteworthy success, there has been a complete failure (witness the cereal leaf beetle, the gypsy moth, and the European corn borer). In general, eradication efforts are most successful when the pest population is relatively small, does not disperse rapidly, and is highly susceptible to control tactics.

Even if a newly established pest population cannot be eradicated, it can often be slowed down a bit by regulating the movement of its host plants or animals, prohibiting sales of infested products, and inspecting all shipments of affected goods. This type of containment strategy usually gives growers and producers a short reprieve before invasion of the pest and buys a little more time for researchers to find better methods of control. By using the containment approach in Texas during a 1971 outbreak of Venezuelan equine encephalitis, entomologists and veterinarians had enough time to reduce populations of mosquito vectors and immunize nearly 3 million horses before the disease could spread into other states.

Licensing and certification are important regulatory tools used in some commodities to ensure that infested or contaminated material is not sold commercially or used as breeding stock. Maintaining disease-free mother plants in strawberries, for example, can often prevent the spread of insect-borne disease during the growing season. Certified seed or plant material is produced under strict guidelines established by commodity organizations in their own interests. These products may be more expensive initially, but eventual savings in pest control usually provide a good return on the investment.

Eugenics and Birth Control

Some control tactics are designed to suppress a pest population by altering its genetic makeup and/or reducing its reproductive potential. As a group, these tactics are frequently known as genetic controls because they affect the accuracy or efficiency with which a pest species passes its genetic material (DNA) from one generation to the next. Genetic control usually works in one of two ways: either by causing (inducing) reproductive sterility, or by incorporating new and potentially deleterious genes (or alleles) into the genetic makeup of a pest population. In effect, some members of a pest species are transformed into biological time bombs that eventually destroy other members of their own species. Because of the self-destructive nature of these tactics, they are sometimes called autocidal control.

Insects can be sterilized by exposing them to certain chemical agents (chemosterilants) or to non-lethal levels of ionizing radiation (X-rays or gamma rays). Chemosterilants are really a form of chemical control. They usually work by blocking the onset of sexual maturity, by inhibiting the production of eggs and/or sperm, or by damaging the chromosomes. These compounds will be covered more fully in the section on semiochemical insecticides.

Exposure to radiation also damages chromosomes (usually by breakage or mutation). Since cells with damaged chromosomes cannot divide correctly, they do not form normal gametes or produce viable offspring. Although the susceptibility of each insect species is different, a proper dose of radiation administered at an appropriate stage of development (usually to pupae) can often induce sterility without causing other deleterious side effects. Sterile individuals, reared in large numbers and released into the environment, can mate with "normal" individuals but they produce no viable offspring. The more sterile individuals released, the fewer "normal" matings are likely to occur.

The effects of radiation on insect development was first studied in 1916, but this knowledge was not put to practical use until the 1930's when E. F. Knipling suggested that it might be possible to suppress pest populations by flooding the environment with large numbers of males that had been rendered infertile by irradiation. By maintaining a constant population of sterile males that was large in comparison to the number of virgin females, Knipling calculated that the number of "normal" matings would decrease each generation until the population was forced into extinction.

The first large-scale test of Knipling's sterile male technique was conducted in 1954 on the island of Curaçao against the screwworm, *Cochliomyia hominovorax*, a Dipteran pest of livestock. After just 13 weeks (four to five generations of the fly) and release of nearly a million sterile males, the screwworm population was completely eradicated from this tiny Caribbean island.

The dramatic success of Knipling's first project led to larger screwworm eradication programs in the southeastern and southwestern United States and to intense interest in the sterile male technique for eradication of other pest species, including Mediterranean fruit flies (*Ceratitis capitata*), boll weevils (*Anthonomus grandis*), horn flies (*Haematobia irritans*), tsetse flies (*Glossina* spp.), pink bollworms (*Pectinophora gossypiella*), and codling moths (*Cydia pomonella*).

Overall, the sterile male technique has had a spotty record of achievement. Despite isolated successes, no other eradication program has ever measured up to the outstanding performance of the screwworm project. From a theoretical standpoint, the use of sterile males is highly attractive because it is non-polluting, species specific, and becomes more effective as the target population decreases. But in practice, the technique simply doesn't work unless a pest population meets the following criteria:

1. *Easy to mass-produce.* A sterile insect release program cannot even be considered unless large numbers of insects can be produced at relatively low cost.
2. *Females mate only once.* This characteristic is uncommon among insects, but it is important for success of the sterile male technique because multiple matings only increase the probability that a female will receive sperm from a fertile male.
3. *Males can be sterilized without loss of competitive vigor.* Sterile males must be able to compete successfully for mates with normal (fertile) males. Sterile males must be physiclly and behaviorally identical to normal males.
4. *Low initial population.* The pest population must be small enough so the initial release of sterile males will outnumber the population of normal males. In some cases, alternative control tactics may be used to reduce initial populations to practical levels.
5. *Restricted geographic range.* Sterile males must disperse to all parts of the range. Small, localized populations are easier to control than large dispersed populations.

Within the last few years, geneticists have begun to devise new genetic control strategies that work by altering DNA or by adding new genes or alleles into the genetic makeup of a pest population. Most of these new tactics are still untested in field applications, but they do offer intriguing possibilities for the future. One such approach involves breeding members of a pest species that have been genetically altered to make them more susceptible to cold winter (or hot summer) temperatures. This type of genetic trait is known as a conditional lethal mutation; it causes mortality only when triggered by extreme environmental conditions. Genetically altered individuals are mass-reared, released into the environment, and allowed to breed with other members of their species. Hopefully, the mutation will spread throughout much of the pest population before the "carriers" are killed by cold (or hot) temperatures.

Another promising approach involves cytoplasmic incompatibility. Although insects lack an immune system, some species do have different "strains" that are not reproductively compatible: the egg cytoplasm of one strain may contain substances that block or inhibit sperm from another strain. By mass-rearing and releasing large numbers of these "incompatible" insects, it may someday be possible to eradicate a pest population and replace it with a more "benign" strain.

Chemical Control

Most people understand that chemical pest control involves the use of chemical substances to kill or disrupt the life cycle of an insect pest. But few people outside the circle of entomology realize just how diverse these compounds are, and in how many different ways they can be used. Although conventional insecticides, the poisons, are still a mainstay of chemical control, they are gradually being superseded by less toxic compounds that disrupt insect development or modify behaviour. Some of these new chemical weapons are much safer for the environment and more species specific than most conventional insecticides.

Semiochemicals — Chemical Control of Behaviour

Much of an insect's behaviour is mediated by chemicals in its environment. By turning these chemicals to our own advantage, it is often possible to attract pests to traps or baits, or repel them from our homes, our crops, or our domestic animals. Behavioral messages are delivered by a wide array of chemical compounds. As a group, these compounds are known as semiochemicals. In some cases, they may facilitate communication between the members of a single species (e.g., pheromones) or between members of different species (e.g., allelochemicals). Functionally, semiochemicals may have a wide range of activity. They may serve as attractants or repellents, they may stimulate or inhibit feeding, they may provoke flight or inhibit it, or they may simply elicit behaviour patterns at inappropriate times.

Attractant pheromones and allelochemicals can be used as lures or baits in a wide variety of insect traps, or they can be mixed together with toxicants to produce an "elixir of death". Protein hydrolysates, for example, serve as feeding attractants for fruit flies (*Rhagoletis* spp.). These chemicals can be applied to sticky traps to improve catch, or combined with an insecticide and sprayed on fruit crops to suppress active infestations. Phenethyl propanoate, eugenol, and geraniol can be mixed in a 3.5:3.5:3 ratio and used as an attractant for Japanese beetles (*Popillia japonica*). These are the active ingredients in the "floral attractant" found in popular bag traps for Japanese beetles. In some cases, chemists have produced synthetic compounds that are even more attractive than naturally occurring chemicals.

Trimedlure, a synthetic substitute for alpha-copaene, is produced commercially as an attractant for the Mediterranean fruit fly (*Ceratitis capitata*). Improved food lures and baits are among the most promising new developments for controlling cockroaches (Blattoidea) in homes and businesses. These are the active ingredients in a new generation of "roach motels" where the insects "check in but don't check out."

Sex pheromones are among the most powerful of chemical attractants. Ever since they were first discovered by A. A. Budenandt in 1959 (from silkworm moths, *Bombyx mori*), these chemicals have aroused great interest because of their potential as pest control agents. During the past 30 years, chemists have identified the sex pheromones for over 300 insect species. Many of these compounds are now sold commercially. In some cases, pheromones are packaged (or encapsulated) in slow-release dispensers (rubber septa, hollow fibres, or rope wicks) that are used as lures in traps of various designs. At low densities, these pheromone traps are a valuable monitoring tool, providing information on the density and distribution of pest populations. At high densities, they can be used for mass trapping sexually active adults (usually males) in efforts to reduce population density and lower a pest's reproductive potential.

Slow-release formulations of sex pheromones can also be used for mating disruption. By increasing the concentration of pheromone in an insect's environment, it may be possible to make everything smell like a prospective mate. Males wear themselves out courting inanimate objects or become habituated to the odor and stop responding to it. This approach, variously known as air permeation or the innundation technique, has shown promise for controlling a number of fruit and vegetable pests, including the codling moth (*Cydia pomonella*), the cabbage looper (*Trichoplusia ni*), the oriental fruit moth (*Grapholita molesta*), and the peachtree borer (*Synanthedon exitiosa*).

Chemical repellents and feeding deterrents are also useful tools for managing insect behaviour. As their name suggests, these compounds cause insects (or other arthropods) to disperse or to discontinue normal feeding behaviour. Repellents such as dimethyl phthalate, benzyl benzoate, and N,N-diethyl-m-toluamide (DEET) have been developed to protect humans from biting flies, ticks, and chiggers (immature trombiculid mites). Other compounds, like di-n-butyl succinate or butoxypolypropylene glycol, are used as fly repellents for cattle.

Moth balls and flakes (paradicholorobenzene or alpha-naphthalene) are placed in drawers and closets to prevent infestation by a variety of insects that feed on stored products and natural fibres. Woolen cloth can be manufactured with colourless dyes (e.g., Eulans and Mitin FF) that bond permanently to the fabric and make it unpalatable to clothes moths and carpet beetles. Wood preservatives, such as pentachlorophenol, act as feeding deterrents to termites and other wood-dwelling insects.

The neem tree, *Azadirachta indica* (Meliaceae) is a promising new source of feeding repellents that may be developed for use on selected non-crop plants. The leaves, twigs, and seeds of this tree, which is grown commercially in India, contain at least 25 biologically active compounds that act as insect repellents, feeding deterrents, or growth regulators. Azadirachtin, the most abundant of these active ingredients, is now commercially available in the United States. Sold under the trade name Margosan-O, this new feeding deterrent has been approved for use on non-food greenhouse crops, ornamentals, and turf.

Chemosterilants — Chemical Control of Reproduction

There are over four hundred chemical substances that are known to cause reproductive sterility in insects. Some of these compounds inhibit ovarian growth and development, while others appear to induce fundamental changes in the chemical structure of nucleic acids (DNA and RNA). These changes (mutations) prevent cell division or obstruct normal embryonic development. Chemosterilants belong to several major chemical groups. These compounds are applied directly to the insect or incorporated into food that serves as a bait. All chemosterilants are extremely hazardous compounds. Their effects are not restricted to insects; they also cause cancer, birth defects, and other mutations in humans and domestic animals. Clearly, these chemicals cannot be dispersed in the environment like other pesticides. Instead, they must be applied under controlled laboratory conditions, usually to insects that are mass reared and released as part of a sterile release program. Although there is much interest in finding a chemosterilant whose effects are limited to insects, no such compound has yet been found.

Insect Growth Regulators — Chemical Control of Development

The enzymes and hormones that regulate developmental processes within an insect's body can sometimes be exploited as chemical control weapons. These compounds, often known as insect growth regulators (IGRs), can be used to stimulate development at inappropriate times or inhibit it at other times. The major groups of IGR compounds include:

1. *Chitin inhibitors.* These chemicals (e.g., diflubenzuron and teflubenzuron) inhibit the molting process (apolysis) by blocking the activity of chitin synthetase, an enzyme needed by epidermal cells when constructing a new exoskeleton. Because of this mode of action, chitin inhibitors are highly specific to arthropods. They act rather slowly (2-5 days), but eventually disrupt any process that involves construction of new cuticle (e.g., molting, hatching, pupation). They are most effective when used against the immature stages of a pest. Diflubenzuron, currently registered under the trade name Dimilin, is used for controlling gypsy moths, boll weevils, and various other pests.
2. *Molting Hormone Analogues.* Ecdysteroids stimulate the molting process by mimicking the action of molting hormone. Applied to the surface of an insect's body or incorporated into its food, these compounds work by initiating premature ecdysis during the immature stages of development. Ecdysteroid-like compounds have been found in some plants where they evidently serve as a defense against insect herbivores. But despite their potential as insect growth regulators, the ecdysteroids have never been developed into commercial products. Their chemical structural is similar to that of human reproductive hormones (estrogen, progesterone, and testosterone), and like many other steroid compounds, they have the potential to cause cancer and birth defects.

3. *Juvenile Hormone Analogues.* Juvenile hormone (JH) and related compounds act as insect growth regulators by inhibiting the developmental changes associated with embryogenesis, morphogenesis, and reproduction. During normal development, JH levels are elevated in larvae (or nymphs) and decrease prior to pupation (or adult eclosion). Contact exposure to JH analogues during the egg stage or after the last larval (or nymphal) molt can inhibit development, delay maturation, and eventually result in death. Since the onset of mortality is usually quite slow (days to weeks), JH analogues have limited utility in agriculture. But several compounds (e.g., hydroprene, kinoprene, and methoprene) have been successfully incorporated into household products for controlling ants, fleas, and other household pests.
4. *Anti-juvenile Hormones.* These unique compounds (the precocenes) were first isolated in 1976 from a common houseplant (*Aegeratum houstonianum*). Precocenes are cytotoxic agents. They become activated by enzymes in the insect's corpora allata, selectively destroying these glands, and preventing all subsequent production of juvenile hormone. In immature insects, exposure to anti-JH compounds may result in premature (precocious) development of adult structures or behaviors. In adults, precocenes can cause sterility because the presence of juvenile hormone is necessary for normal production of eggs and sperm. Anti-JH compounds seem to be most effective against Hemipterans. Despite their unique mode of action, these IGRs have never been developed into commercial products because they break down too rapidly in the presence of oxygen.

Conventional Insecticides — The Killer Chemicals

Conventional insecticides are among the most popular chemical control agents because they are readily available, rapid acting, and highly reliable. A single application may control several different pest species and usually forms a persistent residue that continues to kill insects for hours or even days after application. Because of their convenience and effectiveness, insecticides quickly became standard practice for pest control during the 1960's and 1970's. Overuse, misuse, and abuse of these chemicals have led to widespread criticism of chemical control and, in a few cases, resulted in long-term environmental consequences.

The effectiveness of an insecticide usually depends on when and where the pest encounters it. Most insecticides are absorbed directly through an insect's exoskeleton. These compounds are known as contact poisons because they are effective “on contact”. Other insecticides act as fumigants. They are released in the vapour state (as gases) and enter the insect’s body through its tracheal system. Fumigants are most effective when they are used in an enclosed area such as a greenhouse, a warehouse, or a grain bin. Still other compounds must be ingested before they have an effect. These are known as stomach poisons. They often work more slowly than fumigants or contact poisons, but they are still useful for certain types of pest control in homes and businesses.

Systemic insecticides are a special type of stomach poison. These compounds are absorbed by the tissues of a plant (or animal) without ill effects. Insect pests ingest the insecticide when they feed on the treated organism. Systemic insecticides are sometimes included in the diets of domestic animals to protect them from internal parasites (e.g., cattle grubs and other bot flies). Plant systemics can be incorporated into the soil around ornamentals or bedding plants. The insecticides are absorbed by the roots and translocated to leaves, stems, and flowers. If the insect that feeds on a treated plant doesn’t acquire a lethal dose of insecticide, it may at least be deterred from further feeding. Although systemic insecticides are commonly applied to horticultural plantings, they are not as useful for many food crops because the insecticide remains in the food after harvest.

Insecticides contain one or more active ingredients that serve as toxicants (poisons). In their purest form (technical grade), these chemicals may be too toxic, too unstable, or too volatile to be handled or applied safely. Therefore, technical grade insecticide is always mixed with other compounds, known as adjuvants, in order to improve the performance, safety, or handling characteristics of a commercial product. These mixtures (technical grade insecticide plus adjuvants) are known as formulations. Almost anything could be an adjuvant: pumice, ground walnut shells, buffalo gourd root powder, vegetable oil, etc. These compounds are usually listed on the label as “inert ingredients”, but they are certainly not inactive. Many adjuvants are proprietary products, protected by patents and closely guarded as industrial secrets. They may represent 90-95% of the total volume of a commercial formulation. An insecticide can be manufactured in a variety of formulations, each tailored for specific applications. Some of the more common formulations include:

Granules (G) or pellets (P) — Coarse particles (e.g., clay, ground corn cobs, or walnut shells) can serve as a carrier for certain types of insecticides. The toxicant slowly leaches out of the carrier, minimizing its movement within the ecosystem and maximizing its active life (persistence). Granular formulations are commonly used for controlling soil-dwelling insects and in systemic insecticides that are applied around the base of plants. Since granular formulations do not blow or drift, they are considered relatively safe from the standpoint of accidental human exposure.

Dusts (D) — These dry powders are usually formulated with inert particles of ash, chalk, talc, or clay. They are designed to be sprinkled or blown onto target surfaces. The powder sticks to feet, legs, and other body parts of passing insects, and the toxicant is eventually ingested when the insect cleans itself. Insecticides that act as stomach poisons are often formulated as dusts. These formulations are not always suitable for outdoor applications because they have a tendency to drift (blow away with the wind).

Soluble powders (SP) or wettable powders (WP) — Dry formulations that are mixed in water to form homogeneous spray solutions. A soluble powder dissolves completely in water, whereas a wettable powder contains an emulsifier that produces a uniform suspension (colloid). Many foliar insecticides are formulated as wettable (or soluble) powders because they are easy to transport and generally have a long shelf life.

Emulsifiable concentrates (EC) — These liquid formulations contain the toxicant(s) and an emulsifier dissolved in organic solvent. The concentrate is diluted with a large volume of water to produce the final spray mixture. Some foliar insecticides are formulated as emulsifiable concentrates. Unlike most wettable powders, they do not leave a visible residue on fruits and vegetables. Sensitive plants, however, may be injured by organic solvents in the mixture. Emulsifiable concentrates are also commonly used in sprays for urban and industrial pests.

Aerosols (A) — Insecticides that are formulated together with a solvent may be pre-packaged in pressurized spray cans or sold unpressurized for use in special fogging machines. Spray cans are relatively expensive (per pound of active ingredient) but they are convenient, easy to store, and have a long shelf life. Commercial foggers are typically used indoors (e.g., greenhouses and warehouses) or for control of biting flies in community-wide pest control operations.

Ultralow-Volume Concentrates (ULV) — These highly concentrated formulations (more than 8 pounds of active ingredient per gallon) are designed to be used in specialized spray equipment that atomizes the concentrate droplets. ULV spray equipment is used by most aerial applicators (airplane sprayers) who treat forested lands or large agricultural acreages. Despite their many advantages, conventional insecticides are not ideal pest control agents. Indeed, one of their greatest strengths, broad-spectrum activity, is also one of their greatest weaknesses. While it is certainly an advantage to control multiple pest species with a single chemical treatment, the non-specificity of most conventional insecticides poses a serious threat to non-target organisms in the environment. High mortality among natural enemies can have an enduring impact on the ecological balance of any community. In the absence of biocontrol agents, more insecticide applications may be the only recourse available to stop pest resurgence. Once we step onto this "insecticide treadmill", it can be very difficult to get off.

The non-target effects of insecticides are not limited to natural enemies. Most of the organochlorine compounds (e.g., DDT) have been banned from use in the United States because of their indirect effects on reproduction in birds of prey (e.g., eagles, ospreys, condors, etc.). These environmentally stable compounds are highly soluble in lipids, making it possible for them to accumulate in the body fat of non-target organisms. Predators, particularly those at the top of a food chain, amass pesticide concentrations many times greater than anywhere else in the environment. This process, known as bioaccumulation (or biomagnification), was responsible for high levels of DDT and related compounds in birds of prey during the 1960's and 1970's. These pesticides did not injure the birds themselves, but caused thinning of their egg shells and high rates of breakage during incubation. Most organochlorine insecticides were banned during the 1970's and 1980's and many of the threatened bird species are now recovering from the brink of extinction.

In some cases, sub-lethal concentrations of an insecticide can stimulate rather than suppress the growth of a pest population. This phenomenon, known as hormoligosis, has been observed in a number of pest species, including twospotted spider mites (*Tetranychus urticae*), western corn rootworms (*Diabrotica virgifera*), and brown planthoppers (*Nilaparvata lugens*). Low doses of pesticide seem to improve the nutritional quality of host plants, thereby increasing a pest's reproductive potential or decreasing its time of development.

And finally, there is the problem of resistance to insecticides. Insects are among the most adaptable organisms on the face of the earth. For the past 400 million years they have managed to survive by adjusting to changes in their environment, so it should come as no surprise that they can also adapt to chemical pesticides. Resistance has increased exponentially since the the late 1940's, and today, there are over 500 pest species that exhibit some level of resistance to at least one type of insecticide.

Insects may become resistant to insecticides in several ways. Biochemical resistance usually involves changes in the metabolic pathways that insects normally use to break down plant defenses and other environmental toxins. This detoxication is facilitated by enzymes (esterases, hydrolases, transferases, and oxidases) that change the chemical structure of toxicants before they cause physical harm. Physiological resistance involves functional changes in basic life processes that alter the way toxicants interact with the body. Some German cockroaches (*Blatella germanica*), for example, have become resistant to carbaryl (a carbamate insecticide) as a result of genetic changes in the permeability of their cuticle. Behavioral resistance may occur as a result of any innate change in behaviour that reduces an insect's probability of encountering a toxicant. In some parts of Panama, the mosquitoes that vector malaria (*Anopheles albimanus*) have become hypersensitive to certain insecticides. They manage to escape lethal doses of insecticide by avoiding enclosed areas and refusing to land on treated surfaces.

When an insect population becomes resistant to one insecticide, it may also become less susceptible to other toxicants in the same chemical family. This phenomenon, known as class resistance, is a common problem in all major groups of insecticides (organochlorines, organophosphates, carbamates, and synthetic pyrethroids). When an insecticide loses its effectiveness because of pest resistance, users typically replace it with another compound from a different chemical group. But resistance to one group of compounds does not prevent subsequent development of resistance to compounds in other chemical groups. Over the years, some pests have developed multiple resistance to nearly every insecticide that has ever been used to fight them. Colorado potato beetles, *Leptinotarsa decemlineata*, on Long Island, N.Y. are probably the best example of a population that has developed multiple resistance. It may never be possible to prevent insects from developing resistance, but there are ways to manage these chemicals that will prolong their useful life.

Integrated Control

Just as ground, air, and naval forces are integrated to achieve military objectives, the tactical weapons of pest control can also be integrated to achieve more effective management of pest populations. Development of resistance, effects on non-target organisms, and damage to the environment can all be minimized with selective and judicious use of multi-faceted control tactics. This approach, commonly known as integrated control, requires an understanding of ecological principles as well as a thorough knowledge of the pest's life history and population dynamics. Integrated pest control is not a new concept. It was commonly practiced in the years before synthetic organic insecticides became widely available. But the old ways were largely abandoned after World War II because chemical weapons were so effective, convenient, and inexpensive. Once we recognized the dangers of over-dependence on a single control strategy, the principles of integrated pest control gained renewed acceptance. Today, integrated pest control forms the foundation of Integrated Pest Management programs (IPM) that take a comprehensive and multi-disciplinary approach to solving pest problems. Insects, weeds, plant diseases, and even some vertebrate pests (e.g., birds and rodents) are included under the IPM umbrella. These programs emphasize management rather than eradication. They take a broad ecological approach to pest problems, focusing on all members of a pest complex in an effort to identify the optimum combination of control tactics that will reduce pest populations below economic thresholds and maintain these levels with the least possible impact on the rest of the environment. This approach, often called biorational pest control, relies heavily on cultural and biological tactics that are supplemented with carefully timed applications of highly selective chemical weapons.

The complexity of modern IPM programs will continue to increase as we add more knowledge about pests, develop new management tactics, and learn how to optimize existing control strategies. Mathematical models of population growth rates and plant-pest interactions are being developed and incorporated into computer programs that help us assimilate and interpret data from the many different variables that affect pest population dynamics. These computer programs, usually known as expert systems, are yet another tool in our expanding stockpile of pest control technology.

Integrated pest management is also a major component of Sustainable Agriculture, a holistic approach to modern farming that encompasses not only pest control, but also conservation of soil and water resources, and a wide range of other management practices (e.g., composting, mulching, green manure, waste disposal, etc.). The goal of sustainable agriculture is to satisfy human needs for food and fiber while preserving nonrenewable resources, protecting environmental quality for future generations, and safeguarding the profitability and long-term viability of commercial agriculture.

Chapter

9

Integrated Pest Management

In agriculture, integrated pest management (IPM) is a pest control strategy that uses a variety of complementary strategies including: mechanical devices, physical devices, genetic, biological, cultural management, and chemical management. These methods are done in three stages: prevention, observation, and intervention. It is an ecological approach with a main goal of significantly reducing or eliminating the use of pesticides while at the same time managing pest populations at an acceptable level.

History of IPM

Shortly after World War II, when synthetic insecticides became widely available, entomologists in California developed the concept of "supervised insect control." Around the same time, entomologists in cotton-belt states such as Arkansas were advocating a similar approach. Under this scheme, insect control was "supervised" by qualified entomologists, and insecticide applications were based on conclusions reached from periodic monitoring of pest and natural-enemy populations. This was viewed as an alternative to calendar-based insecticide programs. Supervised control was based on a sound knowledge of the ecology and analysis of projected trends in pest and natural-enemy populations.

Supervised control formed much of the conceptual basis for the "integrated control" that University of California entomologists articulated in the 1950s. Integrated control sought to identify the best mix of chemical and biological controls for a given insect pest. Chemical insecticides were to be used in manner least disruptive to biological control. The term "integrated" was thus synonymous with "compatible." Chemical controls were to be applied only after regular monitoring indicated that a pest population had reached a level (the economic threshold) that required treatment to prevent the population from reaching a level (the economic injury level) at which economic losses would exceed the cost of the artificial control measures. IPM extended the concept of integrated control to all classes of pests and was expanded to include tactics other than just chemical and biological controls. Artificial controls such as pesticides were to be applied as in integrated control, but these now had to be compatible with control tactics for all classes of pests. Other tactics, such as host-plant resistance and cultural manipulations, became part of the IPM arsenal. IPM added the multidisciplinary element, involving entomologists, plant pathologists, nematologists, and weed scientists.

In the United States, IPM was formulated into national policy in February 1972 when President Nixon directed federal agencies to take steps to advance the concept and application of IPM in all relevant sectors. In 1979, President Carter established an interagency IPM Coordinating Committee to ensure development and implementation of IPM practices.

How IPM Works

An IPM regime can be quite simple or sophisticated. Historically, the main focus of IPM programs was on agricultural insect pests. Although originally developed for agricultural pest management, IPM programs are now developed to encompass diseases, weeds, and other pests that interfere with the management objectives of sites such as residential and commercial structures, lawn and turf areas, and home and community gardens.

An IPM system is designed around six basic components: The US Environmental Protection Agency has a useful set of IPM principles.

1. *Acceptable Pest Levels:* The emphasis is on *control*, not *eradication*. IPM holds that wiping out an entire pest population is often impossible, and the attempt can be more costly, environmentally unsafe, and frequently unachievable. IPM programs first work to establish acceptable pest levels, called action thresholds, and apply controls if those thresholds are crossed. These thresholds are pest and site specific, meaning

that it may be acceptable at one site to have a weed such as white clover, but at another site it may not be acceptable. This stops the pest gaining resistance to chemicals produced by the plant or applied to the crops. If many of the pests are killed then any that have resistance to the chemical will rapidly reproduce forming a resistant population. By not killing all the pests there are some un-resistant pests left that will dilute any resistant genes that appear.

2. *Preventive Cultural Practices:* Selecting varieties best for local growing conditions, and maintaining healthy crops, is the first line of defense, together with plant quarantine and 'cultural techniques' such as crop sanitation (*e.g.* removal of diseased plants to prevent spread of infection).
3. *Monitoring:* Regular observation is the cornerstone of IPM. Observation is broken into two steps, first; inspection and second; identification. Visual inspection, insect and spore traps, and other measurement methods and monitoring tools are used to monitor pest levels. Accurate pest identification is critical to a successful IPM program. Record-keeping is essential, as is a thorough knowledge of the behaviour and reproductive cycles of target pests. Since insects are cold-blooded, their physical development is dependent on the temperature of their environment. Many insects have had their development cycles modeled in terms of degree days. Monitor the degree days of an environment to determine when is the optimal time for a specific insect's outbreak.
4. *Mechanical Controls:* Should a pest reach an unacceptable level, mechanical methods are the first options to consider. They include simple hand-picking, erecting insect barriers, using traps, vacuuming, and tillage to disrupt breeding.
5. *Biological Controls:* Natural biological processes and materials can provide control, with minimal environmental impact, and often at low cost. The main focus here is on promoting beneficial insects that eat target pests. Biological insecticides, derived from naturally occurring microorganisms (*e.g.*: *Bt*, entomo-pathogenic fungi and entomopathogenic nematodes), also fit in this category.
6. *Chemical Controls:* Synthetic pesticides are generally only used as required and often only at specific times in a pests life cycle. Many of the newer pesticide groups are derived from plants or naturally occurring substances (*e.g.*: nicotine, pyrethrum and insect juvenile hormone analogues), and further 'biology-based' or 'ecological' techniques are under evaluation.

IPM is applicable to all types of agriculture and sites such as residential and commercial structures, lawn and turf areas, and home and community gardens. Reliance on knowledge, experience, observation, and integration of multiple techniques makes IPM a perfect fit for organic farming (the synthetic chemical option is simply not considered). For large-scale, chemical-based farms, IPM can reduce human and environmental exposure to hazardous chemicals, and potentially lower overall costs of pesticide application material and labour.

1. *Proper Identification of Pest:* What is it? Cases of mistaken identity may result in ineffective actions. If plant damage due to over-watering are mistaken for a fungal infection, a spray may be used needlessly and the plant still dies.
2. *Learn Pest and Host Life Cycle and Biology:* At the time you see a pest, it may be too late to do much about it except maybe spray with a pesticide. Often, there is another stage of the life cycle that is susceptible to preventative actions. For example, weeds reproducing from last year's seed can be prevented with mulches. Also, learning what a pest needs to survive allows you to remove these.
3. *Monitor or Sample Environment for Pest Population:* How many are here? Preventative actions must be taken at the correct time if they are to be effective. For this reason, once you have correctly identified the pest, you begin monitoring BEFORE it becomes a problem. For example, in school cafeterias where roaches may be expected to appear, sticky traps are set out before school starts. Traps are checked at regular intervals so you can see them right away and do something before they get out of hand. Some of the things you might want to monitor about pest populations include: Is the pest present/absent? What is the distribution-all over or only in certain spots? Is the pest population increasing or decreasing?
4. *Establish Action Threshold (economic, health or aesthetic):* How many are too many? In some cases, a certain number of pests can be tolerated. Soybeans are quite tolerant of defoliation, so if you have only a few caterpillars in the field and their population is not increasing dramatically, there is no need to do anything. Conversely, there is a point at which you MUST do something. For the farmer, that point is the one at which the cost of damage by the pest is MORE than the cost of control. This is an economic threshold. Tolerance of pests varies also by whether or not they are a health hazard (low

tolerance) or merely a cosmetic damage (high tolerance in a non-commercial situation). Personal tolerances also vary-many people dislike any insect; some people cannot tolerate dandelions in their yards. Different sites may also have varying requirements based on specific areas. White clover may be perfectly acceptable on the sides of a tee box on a golf course, but unacceptable in the fairway where it could cause confusion in the field of play.

5. Choose an appropriate combination of management tactics For any pest situation, there will be several options to consider. Options include, mechanical or physical control, cultural controls, biological controls and chemical controls. Mechanical or physical controls include picking pests off plants, or using netting or other material to exclude pests such as birds from grapes or rodents from structures. Cultural controls include keeping an area free of conducive conditions by removing or storing waste properly, removing diseased areas of plants properly. Biological controls can be support either through conservation of natural predators or augmentation of natural predators. Augmentative control includes the introduction of naturally occurring predators at either an inundative or inoculative level. An inundative release would be one that seeks to inundate a site with a pest's predator to impact the pest population. An inoculative release would be a smaller number of pest predators to supplement the natural population and provide ongoing control. Chemical controls would include horticultural oils or the application of pesticides such as insecticides and herbicides. A Green Pest Management IPM program would use pesticides derived from plants, such as botanicals, or other naturally occurring materials.
6. *Evaluate Results:* How did it work? Evaluation is often one of the most important steps. This is the process to review an IPM program and the results it generated. Asking the following questions is useful: Did your actions have the desired effect? Was the pest prevented or managed to your satisfaction? Was the method itself satisfactory? Were there any unintended side effects? What will you do in the future for this pest situation? Understanding the effectiveness of the IPM program allows the site manager to make modifications to the IPM plan prior to pests reaching the action threshold and requiring action again.

Integrated Pest Management (IPM) Principles

What is IPM?

Integrated Pest Management (IPM) is an effective and environmentally sensitive approach to pest management that relies on a combination of common-sense practices. IPM programs use current, comprehensive information on the life cycles of pests and their interaction with the environment. This information, in combination with available pest control methods, is used to manage pest damage by the most economical means, and with the least possible hazard to people, property, and the environment. The IPM approach can be applied to both agricultural and non-agricultural settings, such as the home, garden, and workplace. IPM takes advantage of all appropriate pest management options including, but not limited to, the judicious use of pesticides. In contrast, *organic* food production applies many of the same concepts as IPM but limits the use of pesticides to those that are produced from natural sources, as opposed to synthetic chemicals.

How do IPM Programs Work?

IPM is not a single pest control method but, rather, a series of pest management evaluations, decisions and controls. In practicing IPM, growers who are aware of the potential for pest infestation follow a four-tiered approach. The four steps include:

Set Action Thresholds: Before taking any pest control action, IPM first sets an action threshold, a point at which pest populations or environmental conditions indicate that pest control action must be taken. Sighting a single pest does not always mean control is needed. The level at which pests will either become an economic threat is critical to guide future pest control decisions.

Monitor and Identify Pests: Not all insects, weeds, and other living organisms require control. Many organisms are innocuous, and some are even beneficial. IPM programs work to monitor for pests and identify them accurately, so that appropriate control decisions can be made in conjunction with action thresholds. This monitoring and identification removes the possibility that pesticides will be used when they are not really needed or that the wrong kind of pesticide will be used.

Prevention: As a first line of pest control, IPM programs work to manage the crop, lawn, or indoor space to prevent pests from becoming a threat. In an agricultural crop, this may mean using cultural methods, such as rotating between different crops, selecting pest-resistant varieties, and planting pest-free rootstock. These control methods can be very effective and cost-efficient and present little to no risk to people or the environment.

Control: Once monitoring, identification, and action thresholds indicate that pest control is required, and preventive methods are no longer effective or available, IPM programs then evaluate the proper control method both for effectiveness and risk. Effective, less *risky* pest controls are chosen first, including highly targeted chemicals, such as pheromones to disrupt pest mating, or mechanical control, such as trapping or weeding. If further monitoring, identifications and action thresholds indicate that less risky controls are not working, then additional pest control methods would be employed, such as targeted spraying of pesticides. Broadcast spraying of non-specific pesticides is a last resort.

Do Most Growers use IPM?

With these steps, IPM is best described as a continuum. Many, if not most, agricultural growers identify their pests before spraying. A smaller subset of growers use less risky pesticides such as pheromones. All of these growers are on the IPM continuum. The goal is to move growers further along the continuum to using all appropriate IPM techniques.

How do you know if the Food you Buy is Grown Using IPM

In most cases, food grown using IPM practices is not identified in the marketplace like *organic* food. There is no national certification for growers using IPM, as the United States Department of Agriculture has developed for organic foods. Since IPM is a complex pest control process, not merely a series of practices, it is impossible to use one IPM definition for all foods and all areas of the country. Many individual commodity growers, for such crop as potatoes and strawberries, are working to define what IPM means for their crop and region, and IPM-labelled foods are available in limited areas. With definitions, growers could begin to market more of their products as *IPM-Grown*, giving consumers another choice in their food purchases.

If I grow my own fruits and vegetables, can I practice IPM in my garden?

Yes, the same principles used by large farms can be applied to your own garden by following the four-tiered approach outlined above. For more specific information on practicing IPM in your garden, you can contact your state Extension Services for the services of a Master Gardener.

Major Insect Pest Management of Sugarcane Crop

Sugarcane is a long duration crop of 10-18 months and therefore is liable to be attacked by a number of insect pests and diseases. According to an estimate, sugarcane production declines by 20.0 and 19.0 % by insect pests and diseases respectively. To increase the crop productivity, management of insect-pest and diseases is of great significance. Due to diversity in agro-ecological conditions the importance of insect pests and disease varies and therefore, management strategy should be adopted accordingly. Sugarcane is infested by about 288 insects of which nearly two dozen causes heavy losses to the quality as well as quantity of the crop. The scenario of insect pests and diseases varies in sub-tropical and tropical belt of sugarcane. Top borer and stalk borer are found pre-dominantly in sub-tropical areas whereas internodes borer and early shoot borer and among disease rust & eye spot are prevalent in tropical region

Several management strategies have been developed as a result of research and development work. In order to save environment from chemical pollution, use of bio-control has been given utmost attention. The management technologies have been integrated as per need for increasing the efficiency.

Economic Threshold Level (ETL) of Sugarcane Pests

ETL is the density at which control measures should be determined to prevent an increase pest population from reaching economic injury level. This helps to check the pests before they can cause notable damage to sugarcane crop.

Pests of Plants: Sugarcane Stem-borer

Chilo Terrenellus

Profile New Guinea is the 'home' of sugarcane. Commercial sugarcane grown in Australia (and other parts of the world) has been developed from hybrids. These canes are very susceptible to damage by pests occurring in New Guinea and other countries. There are several species of sugarcane stem-borers in New Guinea. The common species in sugarcanes in the coastal villages of the Western Province is *Chilo terrenellus*. Stem-borers cause extensive damage in commercial plantings of sugarcane in New Guinea. The caterpillar feeds within the stem of growing canes and may cause enough damage to kill the growing point, resulting in browning of the upper leaves. Other signs are holes in the stem and frass caught in the axils of the lower leaves.

Some species also feed in the stems of rice, sorghum and corn. Damage to the upper stems of corn plants can result in loss of the male flower followed by poor pollination of the cobs.

Distribution: These pests are widespread in New Guinea and are found intermittently on the nearby islands of Saibai and Dauan in Torres Strait. Damage levels as high as 70-80% have been observed in sugarcane in the Western Province of New Guinea.

Threat: The commercial sugarcane species are softer than the 'wild' canes and have higher sugar content. If these pests were to enter our commercial growing areas they would have a severe impact on production and be very difficult to control.

Keep a Top Watch Don't bring sugarcane from New Guinea into Torres Strait. Don't move sugarcane from the northern islands of Torres Strait. If you see sugarcane with holes tunnelled in the stem report it to your local Quarantine officer. There are also State regulations on the movement of sugarcane. Please observe these regulations to help stop the movement of pests and disease of sugarcane.

Disease Management

- Currently there are about 50 varieties of sugarcane under intensive cultivation in India.
- Over 100 fungi, 10 bacteria, 10 viruses and about 50 species of nematodes are pests of sugarcane in different parts of the world.
- Wherever the crop is grown intensively virulent forms of certain pathogens are chronic.
- When a given pathogen becomes severe enough to get the upper hand, the farmers replace the susceptible variety with resistant or tolerant varieties.
- Thus, in many parts of the world, including India, there are continuous, forced replacements of sugarcane varieties.
- There are over 30 diseases of sugarcane in India, some are of major importance because they are widespread and cause severe losses in yield.
- Red rot is one of the major diseases of sugarcane found in many areas of the world. In India it has caused extensive damage in recent past and remains endemic in severe form in some parts.
- During 1939-1942 it caused heavy losses in Bihar and Uttar Pradesh.

- The disease has been studied in detail during the past half century, as evidenced by the voluminous literature on it.
- Butler and his associates in India investigated the disease in great detail during 1914-1918. This led to the work on breeding for disease resistant varieties to control the disease.
- A lead in this activity was by The Sugarcane Breeding Institute, Coimbatore, South India.

Symptoms

- The first symptom of red rot in the field is discolouration of the young leaves. The margins and tips of the leaves wither and the leaves droop.
- The discolouration and withering continues from the tip to the leaf base until the whole crown withers and the plant dies, within 4 to 8 days.
- In a single stool, most of the stalks may wither almost simultaneously.
- As the disease advances the entire stem rots and the central tissues become pithy.
- The tissues are reddened throughout the basal portion, especially the vascular bundles, which are intensely red; there may be cross-wise white patches interrupting the reddened tissues.
- The internodes may shrink and when such canes are split open large cavities may be found in the centre and the pithy tissues may appear brown.
- Often a profuse whitish growth of the fungal mycelium may be found in the brown background of the host tissue.
- In some cases, black, minute, velvety bodies, representing the acervuli of tire fungus, may also be seen
- Since reddening is a common symptom of other diseases of sugarcane, the white patch symptom is an important diagnostic characteristic of red rot.
- When a diseased plant is open, a characteristic becomes evident.
- In the infected plants the leaves may show symptoms in the form of dark red lesions in the mid-rib, which may elongate, turning blood-red with dark margins and later on with straw-coloured centres.
- In the older lesions minute black dots, representing the acervuli, can be seen. Often the infected leaves may break at the lesions and hand down

- The fungal mycelium is present inter-cellularly, mainly in the parenchymatous cells of the path.
- The hyphae are thin, hyaline, septate, profusely branching and contain oil droplets.
- The acervuli, which are characteristic of the genus, are formed on the surface of the rind, in the leaf mid-rib and sometimes in the pith region. They are minute, black and velvety.
- The compact hymenial layer produces the long, rigid, bristle-like setae, which are septate, dark brown at the base and lighter towards the tip.
- Inter-spaced between the setae are numerous club-shaped to linear, hyaline, single-celled conidiophores, measuring approximately 20 X 8 m.
- The conidia, which are borne singly oil the conidiophores, are single celled, thin walled, hyaline, falcate, granular and guttulate, measuring 16-48 x 4-8m In artificial cultures terminal and intercalary, thick-walled, greenish-black chlamydospores are often produced.
- The conidia germinate readily to produce single germ tubes at the tips of which an elliptical, ovoid or irregular, appresoria are formed.
- The perfect stage of the fungus has been observed both in nature and in culture by several authors.

Disease Cycle

- The planting material, viz., the setts, may harbour the fungus and thus perptuate the disease from season to season.
- The fungus may also persist in the soil oil diseased clumps and dry leaves left in the field after harvest.
- The primary infection, however, appears to be mainly from infected setts.
- Secondary spread in the field may be through irrigation water, cultivation tools anti implements and wind-borne inoculum.
- If the conidia settle on the leaves they may germinate and invade the leaves through various types of wounds including the splitting of the mid-rib so common in many varieties.
- Stem infection may take place through insect bores and root primordia.
- The soil-borne fungus may also enter the healthy setts through cut-ends, and cause early infection of the shoots.

- The prevalence of several pathogenic strains of the fungus has been reported from many countries, including India.
- In general, light-coloured physiologic races sporulate readily and are more pathogenic than the dark-coloured strains that sporulate sparingly.
- Though the perfect stage of the fungus has been observed in nature, the role of ascospores in the disease cycle is not understood.

Control

- Red rot management in sugarcane has become an important issue in all sugarcane areas. While early stage detection may not be quite easy, during later stages, the cane breaks down.
- The canes are to be split open lengthwise to see dull red tissue, interrupted by white patches across the stalk. These patches are characteristic of red rot of sugarcane. Monsoon period enables faster disease spread and drying of the crop.
- The best remedy for avoiding this fungal disease is to cultivate only resistant sugarcane varieties that have been released for cultivation in different sugarcane growing states. Phytosanitation, being the key to manage this disease, stringent domestic quarantine measures to prevent movement of cane setts from endemic areas to new areas has to be enforced by all concerned agencies including sugar mills.
- In recent times, the disease has spread into many states.
- Following good cultural practices such as clearing fields of excessive trash and ensuring efficient drainage.
- Healthy setts only are to be planted to avoid poor plant stand due to rotting.
- Affected fields should be isolated through bunding to prevent movement of water to adjacent fields.
- Rattooning of infected fields should be strictly avoided.
- Crop rotation in the affected fields could reduce disease inoculum.
- Hot water treatment of setts before planting at 52 degree C for 30 minutes is also recommended.
- Planting fungicide treated seed setts is to be followed as a general practice to prevent introduction of the pathogen into sugarcane fields.
- *Chemical Control:* Sett treatment with 0.25% Thiophanate methyl, a systemic fungicide will protect the setts against the primary infection of red rot for 90 days.

- In addition to the various sanitary precautions mentioned above, red rot may also be controlled by growing resistant or tolerant varieties. Inter-generic and inter-specific crosses with Saccharum spp., in the latter case using S.spopntaneum and S. robustum with S. officinarum have yielded many economically important cane varieties, some of which are highly resistant, or at least tolerant, to red rot.

Red-Stripe

Causal Organism: Xanthomonas rubrilineans

Class: Schizomycetes

Order: Pseudomonadales

Family: Pseudomonadaceae

- This is another bacterial disease of sugarcane which has a world-wide distribution.
- It has been known since 1893 in Java and since 1903 in Australia, but in India only since 1933, when it was reported by Desai, and also McRae.
- In 1960 Rangaswami made detailed studies on the disease and its causal agent.

Symptoms

- Red stripe is characterized by the appearance on the leaves of chlorotic lesions carrying dark red stripes 0.5-1.0 mm in breadth and several mm in length, either distributed all over the blade, or concentrated in the middle
- Several of them may coalesce to cover large areas of the leaf blade, and to cause wilting and drying of the leaves.
- Whitish flakes occur on the lower surface of the leaf, corresponding to the red lesions on the upper surface.
- These flakes are the dry bacterial ooze. When young shoots are affected, shoot or top rot may result.
- The growing points of the shoot are yellow and later reddish with dark brown stripes on the shoots.
- The rotting may commence from the tip and spread downwards.
- If the affected plants are cut by splitting the shoot downwards, dark red dis-colouration of the tissues may be seen.
- In the affected canes cavities may form in the pith region, and the vascular bundles are distinct because of the dark red dis-colouration.

- The diseased and rotting shoots can be easily pulled out and separated from the plant.

Disease Cycle

- The disease spreads in the field by wind and rain, and by cutting, as the basal stem from which the setts are taken is mostly free from the bacterial infection.
- The bacterial ooze form the infected leaves and shoots dries up to form a thin crust on the surface.
- When dry, the bacterial cells spread freely by wind.
- The cells falling on the host plants, enter them through natural openings or wounds and establish themselves in the various tissues, including the xylem.
- Infected parenchymatous cells may collapse and normal functioning of the plant parts may fail.
- Several grasses, including ragi and bajra, have been reported to be infected by the bacteria.
- These hosts may also play a role in the perpetuation and spread of the pathogen.

Control

- This is a difficult disease of sugarcane to control.
- When-ever the disease is noticed, the affected plants should be removed and burnt.
- Such systematic destruction of the affected plants reduces the disease incidence.
- Growing resistant varieties is, however,the best method of control.

Sett Rot or Pine Apple Disease

Causal Organism: Ceralocystis paradoxa, Ceratostomella paradoxa

Class: Ascomycetes

Order: Sphaeriales

Family: Ceratostomataceae.

- This disease of sugarcane is common in India and in most of the sugarcane areas of the world including Hawaii, mainland United States, Mexico, West Indies, parts of Africa, and South America.
- In India it is found in all the sugarcane tracts.

Symptoms

- The disease primarily affects the sugarcane setts.
- When diseased setts are planted they may rot before germination, or the shoots may die after reaching a height of about 6-12 inches.
- If infected shoots survive, they are very much stunted and chlorotic.
- Eventually the leaves may wither and the shoots wilt.
- If the affected shoots and setts are examined the central portion of the shoots will be seen discoloured red and the contents of the sett rotting.
- A characteristic pineapple smell is associated with the rotting and hence the name.
- The fungus produces both microconidia and macroconida.
- The microconidia are hyaline when young but become almost black at maturity.
- They are thin walled aμm the chains of microconidia are produced endogenousnd cylindrilorm, measuring 10-15 μm x 3.5-5.0μm in the conidiophores and are pushed out in succession.
- The conidiophores are linear, thin-walled, about 100μm in length and are formed terminally on short basal cells.
- The macroconida are produced in chains of up to 20μm short, lateral conidiophores, which measure 20-80 x 4 μm. They are spherical, elliptical, truncale or sometimes pyriform and are hyaline to olive-green or black, measuring 16-19 x 10-12 μm.
- The fungus produces the ascigerous of perfect stage in culture media and at times on the host tissues.
- The perithecia are gregarious and flask shaped, with a long narrow beak.
- The basal portion measures 200-350 μm in diameter and the beak up to about 1200μm in length and 30-40μm in diameter.
- The asci are clavate, and the ascospores are convex to elliptical and measure 7-10 x 2.5-4 μm.

Disease Cycle

- The fungus is soil-borne, entering the setts through the cut ends.
- Inside the sett it spreads rapidly through the parenchymatoous tissues and causes sett rot.

- The buds on the setts may also be affected and secondary organisms may hasten the decay.
- Occasionally the infected setts may also be responsible for the spread of the pathogen.
- The micro and macroconidia may live saprophytically in soil for several years and infect the setts when planted.
- Insects also play a part in the dissemination of the pathogen.
- The specific role of the ascigerous stage in perpetuating the fungus is not under stood.
- The same fungus also infects banana, papaya, cocoa, coffee, mango, pineapple, arecanut, coconut, palmhyra, date palm and several other hosts.

Control

- Healthy setts should be obtained from disease-free field.
- They should be carefully selected and treated with an organomereurial before planting, to protect the cut-ends from invasion by the fungus.
- Pretreating the setts with hot water has been found to stimulate germination of buds and hasten growth so as to help the young plants to overcome the competition with the pathogen.
- A certain amount of resistance to the disease has been found in a few sugarcane varieties in other countries, but field sanitation practices combined with chemical pre-treatment of the setts are quite effective in controlling the disease.

Wilt

Causal Organism: Cephalosporium sacchari

Class: Deuteromycetes

Order: Moniliales

Family: Moniliacease

- This is one of the early known diseases of sugarcane in India.
- It was first reported by Butler and Khan in 1913, from North India.
- It has been reported to cause severe damage to sugarcane crops in many parts of India. During 1965-1967 it caused severe damage to sugarcane crop in the Deccan plateau
- Same disease has been reported from Mexico, Argentina, Barbados, Columbia, Trinidad, Uganda, South Africa and United States of America.

Symptoms

- The first symptoms of the disease become apparent only when the plant has grown for about 4-5 months.
- The canes show gradual withering.
- On examination of affected clumps, the pith will be seen discoloured purple or dirty reef, with longitudinal streaks
- The leaves of affected clumps gradually turn yellow and dry up.
- A characteristic disagreeable odour is also associated with such diseased canes.
- A cottony white mycelium can also be seen in the pith region.
- Frequently this fungal disease is associated with a saprophytic bacterial growth and often the bacteria are mistaken as causal agents.

Disease Cycle

- The fungal mycelium is abundant in the infected canes.
- The hyphae are hyaline, thin walled and septate.
- They produce numerous microconidia on simple or branched, lateral or terminal hyphae, but NO macroconidia are produced.
- This is an important character which is distint from that of Fusarium. The conidia are oval to elliptical, and measure 4-12 x 2-3μm in size. They are mostly unicellular, but the ones formed later in the advanced growth of the fungus may be septate
- Conidia readily germinate to produce single germ tubes.
- The fungus is transmitted from place to place through the infected seed setts.
- When the diseased setts are planted, the eyes may fail to develop or often the shoots arising from the eyes may wilt, due to the infection spreading to the shoots.
- Root formation in such setts may be very poor. The fungus can also survive in soil as a saprophyte for 2-3 years.
- Near-neutral and alkaline soils are favoured by the fungus. The perfect stage is not known.

Control

1. The disease is controlled by selecting seed setts from disease free areas.

2. Alkaline soils may be avoided for growing the crop. The setts, selected from disease free stalk, should be dipped in organomercurial fungicide before planting.
3. Dipping the setts in 40 ppm of boron or manganese, or spraying the plants with either of these minor elements reduces the disease intensity.
4. The varieties Co-617 and B.P.-17 are more resistant than others; hence they should be used when growing crop in the wilt-sick soils.

Grassy Shoot

- This disease was first observed in Maharashtra on sugarcane variety Co. 419 in 1942.
- At present it is reported to be widespread in parts of Andhra Pradesh, Mysore, Bihar, U.P. and Tamil Nadu.
- The disease is more severe on the ratoon crops than on the first planted ones.

Symptoms

- The disease is characterized by proliferation of vegetative buds from the base of the cane giving rise to crowded bunch of tillers bearing narrow leaves.
- The tillers bear pale yellow to completely chlorotic leaves.
- Cane formation rarely takes place in affected clumps and if formed the canes are thin with short internodes.
- The virus is readily transmitted by sap inoculation and in the field it is transmitted through infected setts and perpetuated through crop ratooning. The aphids are the vectors for this disease.
- The same virus also infects jowar, Napier grass and Madras grass.

Control

- The disease is controlled by eradication of diseased parts as soon as symptoms are noticed
- Avoid selection of setts from diseased area
- pre-treating the healthy setts with hot water at 52°C for 1 hour before planting
- Treating them with hot air at 54°C for 8 hours and spraying twice a month with aphidicides.

Gummosis or Gumming Disease

Causal Organism: Xanthomonas vasculorum

Class: Schizomycetes

Order: Pseudomonadales

Family: Pseudomonadaceae

- This is one of the oldest and most common diseases of sugarcane known in many counties.
- The disease was first reported from Brazil in 1869, since when it has been reported from Puerto Rico, West Indies, Columbia and Australia.
- In India Rangaswami reported its occurrence in Tamil Nadu State in 1960, and it is now known in other sugarcane areas of South India.

Symptoms

- The disease is characterized by three distinct symptoms.
- On the leaves characteristic longitudinal streaks or stripes, 1/8 to ¼ inch in width and several inches in length are found, these stripes are pale yellow in colour, later turning brown.
- The affected canes are stunted with short internodes, giving the plant a bushy appearance. When such canes are split open or cut transversely, a dull yellow bacterial ooze comes from the cut ends.
- In advanced cases, cavities may develop in the pith region.
- The fibro-vascular bundles are deep red and this is more intense at the nodal regions of the stem.
- The causal bacterium is variously named by different authors, but the above name is universally accepted at present.
- The bacterium is a hort rod, 1.0-11.5µm x 0.4-0.5µm, motile by means of a single polar flagellum.
- Gram negative, non-spore forming, non-capsular and non-acid fast.
- On beer extract agar it produces a yellow spreading and slimy growth.

Disease Cycle

- The disease is primarily spread through the setts taken from diseased plants.

- The secondary spread may be through agricultural implements, including the cutting knife.
- In the field the infection may spread by wind and water, the source of pathogen being the gummy bacterial ooze coming out of diseased tissues.
- Certain insects, particularly flies, play a significant role in transmitting the pathogen from place to place.
- The bacterium can survive in the insect's body for a long time and in this way may be transmitted long distances.
- On entry into the host the bacterium reaches the vascular tissues and becomes systemic.
- Besides sugarcane, several plant species falling under Graminae, including maize, sorghum, Panicum spp. and some other grasses have been found to be infected by this bacterium.
- These hosts may help in many ways the bacterial spread and perpetuation.

Control

- As in other diseases of sugarcane, gummosis can be controlled by selecting setts from disease-free plants, and adopting strict field sanitation practices.
- In the Fiji Island the disease was completely eradicated by destroying the diseased canes and growing resistant varieties.
- Studies on the availability of resistant stock among the sugarcane varieties in India are needed.

Foliar Diseases

- Pokkah Boeg caused by Gibberella fujikuroi (Saw.) Wollenw. Wineland with its imperfect stage Fusarium moniliforme Sheldon.
- Downy mildew caused by Sclerospora sacchari Miyake.
- Eye spot caused by Helminthosporium sacchari (Breda de Haan) Butl.
- Yellow spot caused by Cercospora Kopkei Krueg.
- Brown spot caused by Cercospora Longipes Butl.
- Leaf spot caused by Cercospora vaginae Krueg.
- Rust caused by Puccinia kuchnii (Kr.) Butl. and P.erianthi Padwick & Khan.
- Ring spot caused by Letophaeria sacchari Breda de Haan.

- Leaf blotch or spot caused by Helminthosporium halodes Drechsler, H. sacchari and H. tetramera McKinney.
- Root rot caused by Pythium spp.
- Smut caused by Sphacelotheca erianthi.
- Ergot and sugary disease caused by Claviceps spp.
- Root rot caused by Rhizoctonia solani Kuhn and Sclerotium rolfsii Sacc. The symptoms of these diseases are similar to the ones caused by species of the same genus on other hosts.

Root-Knot Nematode

Casual Organism: Meloidogyne javanica

Class: Nematoda

Order: Tylenchida

Family: Heteroderidae

- Over 50 nematode species belonging to 20 different genera are reported to infect sugarcane in various countries of the world.
- At least 10 have been reported in India.
- Among these the Meloidogyme spp. are important.
- Sugarcane is among the various hosts affected by nematodes in India.
- The crop is damaged by the root-knot nematodes, though several other ecto-parasitic nematodes occur on this host.
- The occurrence of root-knot and other hosts was first established by Rangaswami and his associates during 1958-1960.

Symptoms

- The diseased plants are chlorotic and stunted, and yellow stripes show on the young leaves while older leaves appear healthy.
- Crops at all stages of growth are affected and the symptoms are more prominent during summer months.
- When the roots of affected plants are dug out and examined, they are found to be knotted.
- The young, white roots show much less knotting than the older, wiry ones.
- The knots are usually linear and are found more commonly towards the root tips. They are about 5-8 mm in thickness.

- The nematodes are separated from the infected tissues by the Baermann Funnel technique and examined under the microscope.
- The females measure 400-460mm in length and 45-50mm at the thickest point, oesophagus 40-45mm and tail 30-35mm.
- The males measures 420-460mm in length, 20-25mm in width at the thickest point and oesophagus 55-60mm and tail 30-35mm.
- The cyst is pear shaped, measuring 600-650mm x 380-400mm with a prominent beak.
- The other species of Meloidogyne reported on sugarcane are M. arenaria Chitwood and M. incognita Chitwood, both of lesser importance than M. javanica.

Disease Cycle

- The nematode eggs persist in soil in cyst form and when the susceptible host is planted, the cyst give rise to the larval forms which invade the root and form the galls.
- There are reports on the exudation of specific chemical substances by the various hosts with either attract or repulse the nematode larvae.
- Once the larvae are inside the host tissue, they feed through the vascular bundles and complete the life cycle.
- They produce the cysts, which protrude from the host roots and remain there until harvest.
- The production of growth regulators such as indole acetic acid in the host tissue is reported to be correlated with the growth promoting and nodule forming effect of the nematodes, as found by Rangaswami and his associates.
- There are over one hundred plant species, belonging to diversified families, which are host plants for M. javanica. No doubt these host plants, particularly the weeds present in sugarcane fields, and the crops used in rotation with sugarcane, play significant roles in the perpetuation and inoculum build-up of the nematode.

Control

- The root-knot nematode can be checked by fumigating the soil with a nematicide such as Nemagon or D.D.T
- The soil must be opened up soon after crop harvest.
- The chemical fumigant should be injected into the soil as per the directions given by the manufacturer after which the soil

is kept covered with a tarpaulin or plastic sheets for a week or two, if necessary.

- The fumes permeate the soil, killing the nematode larvae and cysts, and other organisms as well.
- The soil is safe for planting in 2-3 weeks after fumigation.
- In some countries crop rotation with marigold reduces the nematode population.
- The effect of such plants or their plant roots exudates on the nematode population, and the inter-relationships of other soil microbial populations with the nematode, have to be investigated in detail to understand this complex phenomena has to evolve effective control measures.
- Certain nematophagous fungi like Catenaria vermicola, Arthrobotrys conoides Drech. and Dactylella deodycoides Drech. are present in soils, and this might help in reducing the nematode population.
- Besides Meloidogyne javanica and other species of Meloidogyne which cause root-knot symptoms, Pratylenchus sp., Paratylenchus macrophallus deMan, Trichodorus sp., Rotylenchus sp., Tylenchorhynchus sp, Xiphinema sp., Rotylenchus sp. and Hoplolaimus coronatus Cobb have been reported by Kishan Singh on sugarcane in some parts of India.
- More detailed studies are needed on the disease symptoms, life cycle of the nematodes, nature and estimation of damage and control measures.
- A phanerogamous parasite affecting the roots of sugarcane has also been reported from some parts of India.
- It is identified as Striga euphrasioides Benth.
- This is a partial root parasite, growing up from the roots to form leafy shoots.
- The parasite can synthesize carbohydrates through the green chlorophyll pigments in the leaves but for its other nutrients it depends on the host root.
- It is usually controlled by pulling out the shoots before flowering and seed set.
- Spraying weedicides like 2-4, D will kill the parasite, without affecting the sugarcane plant appreciably.

Ratoon Stunting

Symptoms

- This is a virus diseases of sugarcane found in some parts of India in recent years.
- The affected plants are stunted, the stunting being most severe in stubble and ratoon crops.
- The setts taken from diseased plants germinate poorly and the few shoots that emerge grow very slowly.
- The virus infects many hosts falling under Graminae.
- It is sap-transmissible and no insect vector has been found.

Control

- The disease spreads mainly through planting material, setts should be selected from healthy plants.
- Treating the seets in hot water at 50^0C for about 2 hours gives 100 per cent control and has become a regular practice in many countries.

Mosaic

Symptoms

- This disease was first observed in Java in 1892 and since then it has been reported from almost every sugarcane growing tract of the world.
- The disease carries different names in different countries but the most common name is 'sugarcane mosaic.'
- During 1919-1920 it caused devastating damage to the sugar industry in Puerto Rico and Cuba, and this led to concentrated studies to establish the causal agent, which was uncertain until about 1920.
- In India the disease was identified during the early thirties of this century and has been reported since then from almost all the sugarcane tracts of the country.
- The intensity of the disease and the loss in crop yield vary considerably, depending upon the host variety and climatic conditions.

Causal Organism

- Sugarcane Mosaic Virus (Marmor sacchari).
- The virus is present in abundance in the chlorotic lesions of the affected leaves.

Chapter

10

Tree Fruit Insect Pests and Diseases

Apple

Apple scab, caused by the fungus *Venturia inaequalis* is a serious disease of apple in British Columbia. It is an annual threat in the wetter interior fruit growing districts (Creston and Kelowna to Salmon Arm) and is even more severe in the coastal areas of B.C. and the Fraser Valley. Apple scab has also caused severe losses in the drier areas of the Okanagan and Similkameen Valleys in years with higher than normal rainfall.

Symptoms

Leaf spots are circular or irregular in shape, and brown to olive green in colour with feathery margins. Severe infections, known as sheet scab, may cause the leaves to fall prematurely. Fruit lesions are circular and brown to black in colour. The skin ruptures around the lesions revealing a dark velvety layer of spores. Older lesions become corky in appearance. The fruit is most susceptible when young. Early infections often result in large fruit lesions and severely cracked and malformed fruit.

Life Cycle

Primary Infection: Apple scab overwinters in infected leaves on the orchard floor. Spores (ascospores) are produced in these dead leaves just about the time that new growth appears in the spring (green tip stage). They are then discharged during rainy periods, and can be carried long distances in air currents. Ascospores lodge on new leaves, blossoms and fruit, and if the leaves or fruit are wet long enough for the spores to germinate, primary infection of fruit and foliage occurs. The primary scab season lasts until all the ascospores from last year's leaves have been released. This varies from year to year but typically ends in late June.

Secondary Infection: Once primary infections have taken place, lesions develop on the leaves and fruit which produce secondary spores (conidia). Conidia are spread short distances mainly by splashing rain, and cause additional spread and build-up of the disease. The time needed for a new infection to begin producing a new crop of spores (incubation period) varies from 9 to 17 days, depending on the temperature. Conidia require a similar or slightly longer period of foliage wetting as needed for a primary spring infection. However, unlike ascospores, they do not require an initial period of rainfall to initiate spore release. Early morning dew will therefore be more important in initiating wet periods that may lead to infection, particularly if followed by showers. Secondary scab lesions can remain active for about a month, producing large numbers of conidia.

Fruit becomes more resistant to apple scab as it matures, requiring increasingly longer wet periods for infection. With prolonged rainy weather, fruit may be infected right up until harvest. Monitoring Primary Infection Periods: The length of time required for infection depends upon the temperature during the time period when the leaves or fruit are wet. The Mills' chart below illustrates the relationship between leaf wetness duration and average temperature, and can be used to determine whether or not an infection period has occurred. The average temperature may be calculated by taking the average of the high and low temperatures during the wet period. When trees are dry less than 10 hours between showers, the periods the leaves are wet should be added together in calculating the infection period.

MacHardy and Gadoury Model for Apple Scab

A new model was developed in 1989 by MacHardy & Gadoury (sometimes called the MacHardy model). This model is based on the fact that ascospore release is strongly influenced by light, and that only a small percentage of ascospores are released at night. Thus, if leaves on the orchard floor are wetted at night by rainfall, the majority of the spores will not be released until the following morning. The

Mills' model assumes that ascospores are released at the beginning of rain. It could therefore overpredict the number of infection periods, assuming that at least some rainy periods every year will start at night. For example, if it started to rain at 2 a.m. and the leaves dried at 11 a.m., the leaves were wet for 9 hours, long enough for a Mills infection period. However, according to the MacHardy model, the length of the infection period was only 3 hours (8 a.m. to 11 a.m.), not long enough to cause an apple scab infection.

Which Model should be Used?

The modified Mills' model has been used successfully in the Okanagan for many years, and continues to be used by most field advisors. There are advantages and disadvantages to both models, and which one is used should be a matter of educated choice. The Mills' model may miss some infection periods that have occurred in shorter wet periods than the chart indicates, and it may overpredict infection periods that start at night. The MacHardy model may not account for rare, massive spore releases that occur at night. (Night-time spore release is more likely to be high following a long dry period, which allows a build-up of mature spores waiting to be released by rain.) Also, in a high inoculum orchard (lots of scab the previous year), the small percentage of spores that are released at night could be enough to cause an unacceptable level of damage. The MacHardy model is recommended in some areas for orchards that have low levels of overwintering apple scab. This may help to save one or more sprays, depending on the season and its weather patterns. If scab was not well controlled last year, it is probably safer to use the Mills model. Comparing the results of both models can be useful, particularly for borderline infection periods.

The suggested threshold used in Ontario for a low ascospore inoculum orchard is less than 0.7% leaves with scab. This assessment should be done in the fall for the following season. Examine the leaves of 600 shoots for scab lesions. (e.g.. 20 shoots on 30 trees or 10 shoots on 60 trees) and record the number of leaves with one or more lesions. Then determine the average number of leaves per shoot. If less than 0.7% of the leaves examined have scab lesions, the orchard can be considered "low inoculum". Be aware that scab may be difficult to identify in the fall, especially when there is leafminer injury present.

Secondary Fruit Infection

Fruit becomes more resistant to apple scab as it matures. Increasingly longer wet periods are necessary for fruit infection as the season progresses. The following tables should be used as guidelines for determining risk of fruit scab.

Apple Scab Control

Resistant Varieties

Commercial apple cultivars vary in their susceptibility to scab infections. Cultivars such as Red Delicious, Golden Delicious, Spartan, Jonagold and Fuji are susceptible, whereas McIntosh, Empire, Mutsu and Gala are rated as very susceptible. A number of apple cultivars are available which are resistant to the apple scab fungus. However, few of these varieties are suitable for long-term storage, or for the commercial fresh fruit market at the present time. Liberty is perhaps the most widely grown scab resistant variety. The resistant variety Goldrush is considered to have commercial potential in some areas. There are many promising resistant varieties under development.

Many resistant varieties are suitable for backyard gardeners, and will eliminate the need for fungicide sprays for apple scab control. Resistant varieties may also be of interest to organic growers or direct farm marketers. Most currently available resistant varieties have a single gene that governs apple scab resistance. As a result, it is not difficult for the apple scab fungus to overcome the resistance (similar to fungicide resistance). Failure of resistant varieties has already occurred in at least one B.C. orchard in the lower mainland, and at other locations in Europe and the United States.

If you are planning to use scab resistant varieties in a commercial setting, it is important to isolate the resistant trees from sources of apple scab inoculum. Constant exposure to apple scab will eventually result in a breakdown of resistance. Consideration should also be given to using fungicide sprays on resistant trees that are exposed to inoculum, at least through the peak period of scab susceptibility in the spring, or at least until scab levels have been brought under control in the orchard. Other cultural control methods, such as leaf removal or leaf shredding will also help to reduce inoculum pressure. Relying solely on apple scab resistance is not a sustainable IPM practice.

Cultural Control

Water Management: Do not run over-tree sprinklers longer than the time needed to cause an infection period. Infection can occur with as little as 6 to 9 hours of leaf wetness at optimum temperatures. The use of over-tree sprinklers can also prolong leaf wetness periods initiated by rain or dew, creating an infection period. When rain ends, shut off over-tree sprinklers until trees have been dry for at least 10 to 12 hours. Do not start the sprinklers when wet weather is predicted.

Orchard Design and Pruning

Good air movement can reduce the number of infection periods because the foliage dries more quickly. An open canopy also improves spray coverage and light penetration. To improve air movement in the orchard, avoid overcrowding and keep trees well pruned.

Reducing Primary Inoculum

Several strategies will help to reduce the amount of overwintering apple scab. Primary inoculum is produced by infected leaf litter on the orchard floor. Any practices that reduce the amount of leaf litter will help to reduce the amount of inoculum. Removing fallen leaves by raking, vacuuming, or using leaf blowers is effective but may not be practical except for small plantings. Home gardeners who remove all fallen leaves from under their apple trees should experience good control of apple scab. Leaves may be safely composted.

Applying nitrogen fertilizers, such as foliar application of urea in autumn, will speed natural decomposition of leaves and help to deactivate the scab fungus. However, the rates of nitrogen required and potential adverse effects of fall nitrogen application on tree vigour and fruit quality have not been fully evaluated. In England, a spray of 5% urea to trees about one week before leaf fall was found to be effective in limiting spore development and hastening leaf decomposition. Research in the United States has also shown large reductions in spore production, often as high as 80 to 90% following application of 5% urea. Care should be taken that urea is not applied too early in the fall. A 5% urea solution may be equally effective when applied to leaves on the orchard floor after they have dropped in the fall. Spring application to the leaf litter has also proved effective in some areas, but it is less likely to have the same impact as fall application. The effectiveness of urea may vary from year to year.

Fall application of dolomitic lime to leaf litter on the orchard floor has been shown to have similar effects to urea application in Oregon, however urea is considered to be more effective. Chopping or shredding the leaf litter in the spring or fall using a flail mower can also dramatically reduce the amount of scab inoculum. Chopped leaves will decompose much faster. Leaf chopping can also be done in combination with urea application.

Note that these control methods do not replace the need for fungicides in commercial orchards. Reducing the initial inoculum will reduce the disease pressure in the orchard, making apple scab easier to control. It will not, however, eliminate the disease, and fungicidal protection will continue to be necessary. Low apple scab pressure is also valuable in helping to prevent or delay the development of fungicide resistance.

Fungicide Resistance Management

The fungicides Nova, Nustar, Sovran, Flint, Vangard and Scala are at risk for resistance. Note, Nova and Nustar belong to the same class (sterol inhibitors), Sovran and Flint belong to the same class (strobilurins), and Vangard and Scala belong to the same class (anilinopyrimidine).

- Apple Fungicide Table-Resistance Management Groups

To help prevent resistance from developing:

- Use recommended tank mixes at label rates
- Alternate between different classes of fungicides. Do not use more than 2 back-to-back sprays of the same class of fungicide.
- Limit use of eradicants to the early primary scab season. Switch to other protectant fungicides after mid-June.
- Maintain good scab control to reduce the probability that a resistant strain may develop.

Apple scab resistance to Senator (formerly known as Easout) and Equal is widespread in the central and north Okanagan, and has been found in other interior fruit growing districts. Because of the extent of resistance, these chemicals are not recommended for scab control. Growers in areas where resistance is known to be low (e.g. south Okanagan and Similkameen), may choose to use these chemicals with caution. Discontinue use if resistance is suspected. Resistance to Senator develops more quickly and easily than resistance to Equal.

Protectant Sprays

Protectant sprays are applied to protect new growth during prolonged rainy periods or in anticipation of a rainy period. The materials provide a film of fungicide which prevents spores from germinating. Protectant sprays are usually applied on a calendar basis, at approximately 7 day intervals. Fungicides with eradicant properties also have protectant properties, and are generally more effective when used in protective programs.

Eradicant Sprays

Eradicant fungicides prevent or reduce further growth of infections that have already started, if the sprays are applied soon after the start of a rainy period. The advantage of eradicant sprays is that they are applied only when necessary; i.e. when infection periods occur. In most seasons this will reduce the number of sprays required. However, growers must have spray machinery available to cover the trees quickly after an infection period, and winds must be low enough to permit good coverage. Nova, Nustar, Flint and Sovran may be applied up to 96 hours after the start of a rainy period, while Vangard and Scala must be applied within 48 hours after the start of a rainy period for eradicant activity to be effective.

Spray Timing

First Spray: The risk of apple scab infection may begin as early as the green tip stage. If using a protectant program, and scab was a severe problem the previous year, the first spray should be applied at the green tip stage. If scab was very well controlled last year, it may be safe to delay the first spray until the tight cluster stage. Sprays may also be delayed if there is no possibility of rain in the forecast. If using an eradicant program, the first spray should be applied immediately after the first infection period. Begin monitoring infection periods at the green tip to tight cluster stage, depending on severity of scab the previous year.

To successfully use an eradicant schedule, the grower must be able to determine whether or not an infection period has occurred after every rainfall. Many packinghouses issue scab warnings using code-a-phones. Alternatively, growers could install weather monitoring equipment in their own orchards, or hire orchard consultants who are monitoring infection periods.

Scheduling: Scheduling: Sprays may be applied either on a protectant or an eradicant schedule. For a protectant schedule, fungicides should be applied every 7 to 14 days during the primary scab season. A 7 day interval should be used from the pink to calyx stage of tree development. Following the calyx stage, protectants such as Kumulus are best used on a 7 day interval, while Polyram, mancozeb, or captan may be used at 10 to 14 day intervals during most years. This interval should be shortened to 7 days following heavy rain. With an eradicant schedule, fungicides are applied shortly after an infection period has occurred. Often protectant and eradicant schedules are combined. For example, eradicant schedules may be backed up with protectant sprays during the critical pink to petal fall period. Refer to the Fungicide Notes section for information on individual products.

A protectant schedule is highly recommended during the critical pink to petal fall period. During this time there is abundant scab inoculum, the weather is often rainy, and tree growth is rapid and highly susceptible. Protectant schedules are also recommended for orchards with high levels of overwintering apple scab inoculum.

Fungicide Notes

Nova is an eradicant fungicide with 4 days of kick back, or post-infection activity. For bloom and post bloom sprays, tank mix the full rate of Nova with the half rate of a suitable protectant such as Dithane, Polyram or captan. Tank mixing Nova with a protectant will improve scab control on the fruit and will also help to prevent resistance problems from developing. Nova + half rate protectant is recommended for use on a 10-day protectant schedule. If the first application of Nova was applied as an eradicant (post infection), then a second application of Nova or Nova plus protectant should be applied in 7 days. This will help to ensure complete eradication, particularly under severe scab conditions. If switching from Nova to a straight protectant program (e.g. full rate of captan, Dithane or Polyram), the full rate of protectant should be applied no later than 4 to 5 days after the last Nova spray. This reduced spray interval is necessary because the half rate of protectant used with Nova may not last beyond 4 to 5 days. Control failures have occurred in the Okanagan where this spray interval was stretched beyond 5 days.

Nustar belongs to the same class of fungicides as Nova, and also provides 4 days of kick-back. Nustar may be used on a protectant schedule at 7 to 10 day intervals. Use the shorter interval on susceptible varieties or when rainy weather results in frequent infection periods. Nustar should be used in a tank mix with Manzate or captan, as per label rates. Use the high rate of Nustar if powdery mildew control is also needed. Do not apply more than 4 times per season or within 77 days of harvest.

Sovran: Apply as a protectant beginning at inch green or when conditions become favourable for primary infection. Continue on a 10-14 day interval, using a 10 day interval during periods of rapid shoot growth. Sovran provides 96 hours of kick-back for apple scab control. Use the high rate if applying Sovran post-infection or under high disease pressure. Do not apply more than 4 times per season or within

30 days of harvest. Alternate with fungicides of other classes for resistance management. Caution: Sovran drift may severely injure cherries, resulting in leaf burn and defoliation.

Flint belongs to the same class of fungicide as Sovran, and also provides 96 hours of kick back for apple scab control. Flint may be used on a protectant schedule at 7-10 day intervals. Do not apply more than 4 times per season or within 14 days of harvest. Alternate with fungicides of other classes for resistance management. Vangard is best used as a protectant fungicide, however it does have 48 hours of kick-back for apple scab control. Protectant sprays should be applied at 7 to 10 day intervals. Vangard is recommended at the full label rate for pre-bloom sprays. Do not apply more than 2 applications of Vangard per season, and do not apply within 72 days of harvest. Tank mixes using a reduced rate of Vangard have not been fully evaluated in B.C. Caution: Vangard drift may injure some cherry varieties.

Scala is best used as an early season protectant fungicide, and has good activity during cool temperatures. Apply at 7-12 day intervals, from green tip to petal drop. Do not apply within 72 days of harvest. Scala provides better scab protection on leaves than on fruit, thus it is not recommended for scab control after bloom.

Secondary Scab Season

In late June growers should carefully check leaves and fruit for the presence of scab lesions. If primary scab has been well controlled, such that only a few lesions can be found per tree, the interval between sprays may be lengthened for the remainder of the growing season. If scab lesions are not found, the grower could decide to eliminate fungicide applications entirely after the end of the primary season. However, this should be done with caution because it takes only a small number of lesions to produce a major epidemic later in the season if leaves and fruit are not protected, and if an extended period of wet weather occurs. If sprays are suspended, the orchard should be periodically monitored for the development of scab lesions for the rest of the season.

Pin-Point Scab and Storage Scab

Pin-point scab is simply another name for late season apple scab fruit infection. As fruit matures, it becomes increasingly resistant to scab. However, infection can occur at any time up until harvest during prolonged wet weather. Late season infections may not become visible until the apples are in storage. They appear as either pin-point sized lesions, or jet black spots. Fortunately, scab will not spread from diseased to healthy apples in storage.

Pin-point or late season scab infections can develop only when there are visible leaf or fruit infections in the tree. For this reason, it is important that spray programs be continued throughout the summer if scab is present in the orchard. In the Okanagan, spray programs are often relaxed during dry summer weather. Be aware that this may result in fruit infection if wet weather occurs while the trees are unprotected and scab lesions are present in the orchard. Captan and ziram are used as late season protectants. Do not use eradicant fungicides as protectants for pin-point scab, and observe pre-harvest intervals for all late-season sprays.

Ascospore Monitoring

Ascospore monitoring may be used as a technique to more accurately predict the beginning, the peak, and the end of primary apple scab season. Infected leaves are collected and examined using a microscope to determine the stage of spore development. It may be possible to eliminate one or two early season sprays if ascospore monitoring indicates that the scab spores are not yet mature. Similarly, unnecessary sprays at the end of primary scab season could be eliminated if the majority of spores have been discharged earlier than anticipated. There is no valley-wide program at the current time for ascospore monitoring. A limited number of locations are monitored in some years by BCMAL and Agriculture & Agri-Food Canada for research purposes. In the absence of monitoring data, it is assumed that spore release begins at green tip, peaks around blossom, and finishes by mid to late June. Monitoring to date has indicated that these guidelines are very close to reality.

Action Thresholds

Recently, there have been techniques developed that can assess the "potential ascospore dose", or PAD in an orchard. If the PAD is determined to be low enough, it is deemed safe to delay the beginning of the spray program until the pink fruit bud stage or until after 3 infection periods, whichever comes first. To determine PAD it is necessary to do an assessment of scab levels on leaves in the fall. The use of PAD as a threshold has not been evaluated in B.C.

Weather Monitoring Equipment

To forecast infection periods and properly time fungicide applications it is necessary to have access to weather data. Okanagan packinghouses operate a network of electronic weather stations. Downloaded weather data is run through computerized apple scab models to determine when infection periods have occurred. Growers can call into their packinghouse code-a-phones to find out the latest recommendations for scab control. Because microclimates differ between orchards, it is always preferable to collect weather data as close to your own orchard as possible. Growers with larger acreages of apples may consider purchasing their own weather stations. There are several on the market that include software for forecasting apple scab. For more information on weather stations, contact your field advisor.

Bacterial Canker of Stone Fruits

Bacterial canker has been an increasing problem in the British Columbia interior, particularly on young cherry trees. The disease is caused by the bacterium *Pseudomonas syringae*. Cherries are very susceptible, but *Psuedomonas* also infects other stone fruit, pears, apple rootstocks and many species of ornamental trees. Young cherry trees are more susceptible than established cherry trees, and trees under stress are much more susceptible than healthy trees with optimal growing conditions. Outbreaks are often associated with prolonged periods of cold, wet weather with late spring frosts.

Symptoms: Symptoms on young cherry trees are typically elongated cankers that are soft or spongy to the touch and gumming copiously. Cankers may expand rapidly in the spring causing girdling of the main trunk or branches. Bacterial canker can also kill buds, and sometimes causes brown, circular lesions on leaves which fall out to give a "shothole" type symptom. Fruit lesions include small, brown spots which may be slightly sunken on immature fruit. Leaf and fruit symptoms are not common in the Okanagan, but may be seen in areas with higher rainfall.

Laboratory diagnosis should be done in the spring, as the bacterium is difficult to isolate once hot summer weather arrives. Bacterial canker is similar in appearance to Cytospora canker.

Life Cycle: Bacterial canker infections are thought to take place in the fall and winter during cool, wet weather. Trees are particularly susceptible during autumn leaf fall, when fresh leaf scars may become infected. Cankers may not be obvious until the spring, when they start to expand rapidly. Frost damage in the spring may promote additional infections. The bacteria overwinter in canker margins, in healthy buds and also systemically in the vascular system. In the spring bacteria are disseminated by rain to blossoms and young leaves. The bacteria can survive in an epiphytic phase on the surface of symptomless leaves and blossoms, and also on other plants or weeds in the orchard during the summer.

Control:

1. Minimize stress on young or recently planted cherry trees, as it is a major predisposing factor. For example, provide adequate water to prevent drought stress; avoid planting in areas with poor drainage or high frost potential; supply optimal levels of nutrients; check soil pH; and control other pest and disease problems that may weaken trees.
2. Don't let weeds grow up around the trees. They cause unnecessary stress by competing for moisture and nutrients, increase humidity, and may harbour populations of the bacteria that cause canker.
3. Remove severely affected trees, and prune off dead or dying branches. Pruning during dry weather in mid-summer is unlikely to spread the disease, but disinfecting tools after cutting infected trees is probably a good precaution. Avoid pruning in early spring and fall when bacteria are most active.
4. Trees with minor gumming may recover. Cankers may heal if the tree is able to contain the canker with callus tissue. Small cankers may also be cut out with a sharp knife, being careful to disinfect cutting tools between trees.
5. Applications of copper oxychloride (fixed copper) in the fall to protect leaf scars, and in early spring before bloom can be helpful, however copper resistance is widespread in the Pacific Northwest in orchards and nurseries with a history of copper use. It is not known whether copper-resistant strains are present in B.C.
6. The F12-1 Mazzard rootstock is reported to be resistant. If propagating cherry, use only scions from virus-free, canker-free trees.
7. Before establishing new orchards, have soil tested for pathogenic nematodes. High nematode populations, particularly the 'ring nematode' are associated with increased losses due to bacterial canker. Fumigate soil before planting if nematode populations are high.

Bull's Eye Rot of Apple and Pear

Bull's eye rot is a disease that typically appears on apples during storage. Symptoms are usually not present at the time of picking. The disease is caused by the same fungi that cause perennial and anthracnose cankers on the limbs and trunks of trees. Spores may be spread from cankers to fruit any time during the growing season, but the risk of infection is probably greatest close to harvest.

Symptoms: Fruit lesions are circular, brown and sunken with a target appearance due to concentric of gray to cream coloured fruiting bodies.

Control

Cultural Control: Prune out and remove from the orchard as many cankers as possible during winter pruning. Remove any new cankers that develop on limbs and trunks during April through July. Remove as soon as they appear any 1-year-old terminal shoots that wilt or die suddenly during April through July; these shoots are probably girdled by canker.

Chemical Control: Apply calcium sprays to improve fruit resistance to disease. A pre-harvest spray with ziram for pinpoint scab may also help to control Bulls eye rot. Good cultural practices will reduce or eliminate the need for chemical control.

Calcium Deficiency (Bitter Pit) in Apple

Calcium deficiencies contribute to certain fruit disorders including bitter pit, internal breakdown, storage rots, cork spot (Anjou pit), and alfalfa greening. Jonagold and Golden Delicious are particularly sensitive to bitter pit development. The need for calcium sprays is not predictable from soil or leaf analysis. Calcium chloride sprays alone can not be expected to eliminate calcium related fruit disorders. Cultural factors can influence calcium nutrition.

Bitter Pit of Apple, Cork Spot (Anjou Pit) and Alfalfa Greening of Pears

Bitter pit on apples shows as small pits that resemble miniature bruises. The skin within the pit is grey, brown or black and the flesh beneath is dry, brown and spongy. Bitter pit is most common on vigorous trees bearing light crops and is especially common on large fruit. The symptoms may show in the orchard but more frequently they become apparent after picking. Bitter pit is a disorder that results from an inadequate supply of calcium to developing fruit. Cork spot, often called Anjou pit, affects only Anjou pears. Pears develop a bumpy, uneven surface as they approach maturity. Affected areas turn yellow and soften prematurely. The flesh beneath the depressions is soft and turns brown or grey. Anjou pit occurs most seriously in seasons when hot, dry weather precedes picking. Cork spot is a disorder that results from inadequate supply of calcium to developing fruit.

Alfalfa greening (Green stain) affects only the Anjou variety of pears. Fruits develop dark green, slightly sunken areas on the skin shortly before picking. Alfalfa greening is a physiological disorder, most likely caused by a nutritional imbalance between calcium and nitrogen, and excessive irrigation.

Control of Calcium Related Disorders

Cultural

1. Avoid dormant pruning practices that stimulate tree vigour.
2. Reduce nitrogen application to moderate tree vigour.
3. Maintain a regular but not excessive irrigation supply.
4. Practice summer pruning in vigorous, light crop trees in mid-July.
5. Lime when soil pH is low.

Chemical

Calcium chloride sprays can be expected to reduce the incidence of many calcium deficiency disorders in fruit.

Apples: Apply calcium chloride (food grade, 27% calcium) sprays at 8-12 kg/ha (3.3-4.9 kg/acre) or 500 g/100 L (2.27 kg/100 gal) at 10 day intervals beginning mid July. Lower rates cause less leaf burning but should be applied more frequently to be similarly effective. Aim for a total application rate of 36 kg/ha (15 kg/acre) per orchard per year.

Apply COR-CLEAR (34.5% calcium) at 6.3-9.4 kg/ha (2.6-3.8 kg/acre) in 500-2500 L/ha (200-1000L/acre) of water for concentrate applications. For dilute applications, use 0.4kg/100L of water. Aim for a total season application rate of at least 28 kg/ha (12kg/acre).

Warning: Application of calcium chloride almost always causes non-serious burn of leaf edges but severe leaf and fruit injury is possible under some conditions. To reduce the chance of severe injury, be careful not to exceed recommended rates. Do not apply the spray under wet, humid, or slow drying conditions. Spraying when temperatures are high may also produce injury. Although the safe upper temperature limit is not known, it is suggested that damage may occur at about 27 °C. In hot weather, growers should apply the spray during the coolest part of the day.

Pears: Calcium chloride (food grade, 27% calcium) should be applied as a high volume spray of 400 g/100 L (1.8 kg/100 gal) with a surfactant. Do not apply at the apple rate as this may result in leaf and/or fruit burn. Apply three or more sprays at 2-week intervals beginning mid-May. Apply COR-CLEAR (34.5% calcium). For dilute applications use 0.4 kg/450L of water. Apply COR-CLEAR at up to 4.5 kg/ha (1.8 kg/acre). For specific recommendations, consult your fieldman or Horticultural consultant. For more information consult the BCMAFF Tree Fruit Production Guide.

Crown Rot

Causal Agent: Phytophthora Cactorum

Although crown rot is primarily a disease of apples in British Columbia, it has been identified on peaches, plums, apricots and cherries.

Symptoms: In apples, the first sign of disease is usually a purple-red colouring of leaves in late August and early September. The following year, infected trees are either dead or unthrifty with pale, sparse leaves. The inner bark of the crown and main roots of a diseased tree is orange to red-brown. In time the diseased bark becomes soft and dark brown in contrast to white bark in a healthy tree. On some trees, only the roots and not the crown show symptoms. Note that leaves on trees with M9 and M26 rootstock may turn red in late September and October without being infected with crown rot.

Life Cycle: Crown and root rot is caused by the soil-borne fungus, Phytophthora. This fungus requires wet soil for spore production and tree infection. Trees between the ages of three and ten are most commonly affected.

Cultural Control

1. Select rootstock for all new apple plantings from among the more resistant types. Losses have been less frequent on M4, M9 and MM111. In recent plantings of M26, crown rot losses have been more common than in past years. These losses may be due to winter injury predisposing the rootstock to crown rot. Intermediate susceptibility has been observed on M2 and seedling rootstocks. Losses have been most frequent on M7, MM104 and MM106, with MM104 and MM106 being the most susceptible.

2. Inarching partially girdled mature trees with resistant seedlings may overcome the damage from infection.
3. To stop enlargement of the lesions on partially girdled trees, expose the crown, scrape away infected bark and air dry. Replace earth around trunks before winter.
4. To reduce the risk of collar rot in wet soils, plant apple trees with the graft union above, but no more than 5 cm above, the soil surface.
5. Avoid prolonged soil saturation or standing water around the base of the tree.
6. Control weeds around the trunk of apple trees. They can serve as alternate hosts for the crown rot fungus.
7. Reduce nitrogen application on trees with excessive growth as they are more susceptible to crown rot.

Chemical Control

Bearing Apple Trees

Apply Aliette WDG (fosetyl al) as a foliar spray during the tight cluster or pink stage when there is sufficient leaf area present to take up the spray. Use a maximum rate of 5.0 kg/ha (2.0 kg/acre) in a maximum volume of 1000 L/ha (400 L/acre) on standard sized trees. Adjust the spray volume according to the tree row density and canopy (Tree Row Volume). Repeat approximately 6 weeks later. Treat again in the fall soon after harvest. Do not apply more than 3 sprays (15 kg Aliette/ha) per year, and do not treat within 30 days before harvest. Aliette should be sprayed to wet, not to run-off. Do not use a drench treatment for the application of Aliette on bearing trees.

Non-bearing Apple Trees

Apply Aliette WDG as a high volume coarse spray (5 L of water per tree) using a handgun to drench the trunk and soil surrounding the tree. The application rate varies from 5 to 10 grams per tree, to a maximum dose of 5.0 kg/ha (2 kg/acre). Make the first application in spring any time after silver tip and again in early fall. Do not apply as a drench more often than twice per year. Aliette may also be applied to non-bearing trees as a foliar spray if there is sufficient foliage present for good uptake of the chemical.

Apply Ridomil Gold 480 EC (metalaxyl-m) at the rate of 1.0 mL in 5 litres of water per tree, thoroughly drenching the soil around the base of the tree. The first application should be made at the time of planting (preventative treatment), followed by a second application in late August. Treatments can be repeated in the spring (before new growth begins) and late August, in each subsequent non-bearing year depending on the occurrence of crown rot in the orchard. Do not mix Ridomil with herbicides.

Caution-Resistance to Ridomil has already been observed for related species of crown rot fungus. Indiscriminate use of Ridomil may result in the development of resistant crown rot fungus. Use only when necessary.

Cytospora Canker (Leucostoma Canker)

Causal Agent: Leucostoma Cincta

Cytospora canker is an important problem on stone-fruit trees, especially in the South Okanagan and Similkameen. The majority of infections are found on scaffold limbs or trunks of infected trees. Both pruning wounds and winter injury are major points of entry of infection.

Symptoms include the presence of dead twigs or branches after the tree has leafed out in the spring. Closer examination of dead limbs often reveals slightly sunken areas in the bark. Small, black, pimple-like fruiting bodies of the fungus often develop under the bark in these sunken areas. Later in the spring, hair-like masses of spores are extruded from these pimple-like structures. Conidia (spores) are most abundant in the fall and spring. During rain or irrigation, spores are splashed and blown around the orchard. Infection occurs through injuries to the bark such as pruning wounds, leaf scars, winter injury, and sunburn. After the fruiting bodies have been washed by numerous rains, small white dots remain on top of the pimple-like structures and serve as useful indicators of the fungus. Young cytospora cankers can be difficult to distinguish from bacterial canker.

Recent research in Washington found that spores were produced throughout the year and were present in highest numbers during spring and summer on both cherry and peach. Spore production and dispersal resulted from the use of over-head and under-tree impact sprinklers. It was concluded that disease management could be improved by alteration of horticultural and irrigation practices.

Cultural Control

1. Trees should be maintained in a vigorous state.
2. Prune during mid- to late-winter when inoculum levels are lower.

3. To encourage rapid healing-over, cut branches just beyond the ridge of the thickened bark which connects them to larger limbs. Do not leave pruning stubs. Avoid large pruning cuts.
4. Trees should be trained properly so that wide crotch angles are developed.
5. Summer pruning results in many wounds that can be infected by cytospora canker. Summer prune in dry weather and allow time for pruning wounds to heal before using overhead sprinklers.
6. Sporulating infections on scaffold limbs or trees should be removed immediately and burned because they are a source of spores.
7. Use preventative measures to minimize winter injury, sunburn, rodent damage and insect damage.
8. Do not plant new blocks of stone fruit next to old cankered blocks. Nursery stock should be canker free and protected from wood boring insects such as peach tree borer.
9. Consider the effects of irrigation systems on disease development when planning new orchards. E.g. Drip or microsprinkler irrigation would result in less disease pressure than overheads.

Chemical Control

1. There are no fungicides registered in Canada for the control of cytospora canker on stone fruit.

European Canker (Nectria Canker)

Causal Agent: Nectria Galligena

European canker is severe in coastal orchards, and has been a problem in some interior orchards in high rainfall areas. The disease has been introduced to orchards on nursery stock grown in wet climate areas. At the time of purchase infected nursery stock may have no obvious symptoms of canker, but cankers develop during the first year after planting. While the disease is less aggressive in the dry climate of the Okanagan/Similkameen valleys, natural spread does occur and it is capable of causing considerable damage. Spores produced in cankers may also cause a fruit rot similar to bulls eye rot.

Symptoms

The first sign of canker is a reddish-brown spot around a leaf scar, spur or pruning wound, usually in late spring or early summer. The canker develops quickly with ring shaped cracks forming in the bark, which may appear swollen. If the canker girdles the trunk or branch, shoots above the canker die. The canker usually stops growing after one year. A ridge or collar of healthy tissue walls off the canker. Young cankers produce small, whitish fruiting bodies. Older cankers may produce orange-red fruiting bodies.

European canker can sometimes be confused with anthracnose canker.

Life Cycle

Spores from new cankers are spread primarily by rain or overhead irrigation. New infections are thought to occur mainly in the autumn through leaf scars, with symptoms appearing the following spring. Older cankers produce airborne spores (ascospores) that can initiate new infections at a distance from the source. The airborne spores function mainly to initiate new infections, while the water-borne spores serve to intensify the disease in trees that are already infected.

Cultural Control

1. Prune out diseased wood. On young trees this often means pulling out the tree or pruning it back severely.
2. Nursery trees should be examined carefully for symptoms of the disease at planting and again the following spring. Trees with cankers should be returned to the nursery for replacement or discarded.

Chemical Control

1. There are no fungicides registered for control of European canker in Canada. Fungicides applied for other diseases such as apple scab may help to reduce the numbers of new infections. Copper fungicides such as Bordeaux mixture are recommended in some wet climate areas.

Fire Blight of Apple and Pear

What is Fire Blight?

Fire blight is a destructive bacterial disease of apple and pear, recognized by a severe blighting of blossoms, shoots, limbs and fruit. It also affects several ornamentals in the same family such as hawthorn, mountain ash, cotoneaster, quince and firethorn. Outbreaks of fire blight occur periodically in British Columbia pear orchards. Fire blight has also been an increasing problem in high density apple plantings of newer, more susceptible varieties.

Symptoms: Fire blight symptoms may appear on the blossoms, shoots, branches, trunk and rootstock. Blighted blossoms appear wilted, shriveled and brown. Young fruitlets are also very susceptible and appear water soaked and slightly off-colour soon after infection. Fruitlets quickly turn brown to black and eventually shrivel up.

Blighted pear shoots are black in colour, while infected apple shoots are usually a lighter shade of brown. Infected shoots (or "strikes") wilt rapidly, and often form a shepherd's crook at their tips. During warm and humid or rainy weather drops of milky to amber coloured bacterial ooze frequently appear on the blighted shoots and fruit. Blighted leaves may remain attached to the tree throughout the winter. When shoots attached to scaffold limbs or trunks are attacked, the pathogen may spread into the structural wood causing cankers. In susceptible hosts or young trees the disease may travel rapidly down branches causing girdling and death of the branches or sometimes the main trunk.

Cankers appear as slightly darker, water soaked areas in the wood, which may produce amber coloured bacterial ooze that runs down the bark. Reddish brown streaks may be seen in the cambium under the bark of diseased branches. Later in the season the bark often cracks around the margins of the canker. Shoot blight caused by reactivating cankers in the spring is known as canker blight. Bacteria that overwintered in holdover cankers begin to multiply and invade nearby shoots or water sprouts. Invaded shoots wilt and die, and may be mistaken for early shoot blight. Reactivating cankers may also expand to girdle limbs.

Fire blight may also spread into the root area, leading to tree death. Rootstock blight may not exhibit typical fire blight symptoms. It is sometimes mistaken for crown rot. Apple nursery stock with fire blight infection in the rootstock was observed in 1994 in several Okanagan orchards. Infected trees weakened and died beginning 1 to 3 months after planting. Oozing often occurred near the base of the tree. Cutting into the crown area below the bud union, or in some cases the roots, revealed dead, brown tissue. In some cases, leaves appeared yellow or reddish in colour early in the season — a general symptom of root problems.

Life Cycle: Fire blight bacteria overwinter primarily in cankers on infected trees. During spring and early summer, cankers that were not removed the previous season may reactivate and produce bacterial ooze, which may or may not be visible. This ooze, consisting of millions of bacterial cells, is easily transported to blossoms by insects such as flies, ants, and beetles. If weather conditions are warm and humid, the bacteria are able to multiply rapidly in the blossom nectar. Bacteria can then be spread very efficiently from blossom to blossom by honey bees. This gives fire blight explosive potential if the conditions for blossom infection are met.

Fire blight blossom infection is favoured by moderately high temperatures and rainy or humid weather. Temperatures of 18°C and above favour rapid infection whenever moisture or dew is present. Temperatures below 15.5°C retard blight development. In the Okanagan, prevailing cool temperatures during the main blossom period are normally not favourable for infection. Infection is more likely to take place during secondary bloom when temperatures are higher. Once blossoms are infected, the bacteria can quickly spread into shoots and branches. Blossom infections become visible as "strikes", or dying shoots in anywhere from 1 to 4 weeks, depending on the temperature.

Infected shoots provide additional sources of fire blight bacteria, which can be spread by rain, (especially wind-blown rain), insects, and contaminated pruning tools. Secondary infections may continue to occur throughout the growing season. The worst epidemics always follow blossom infection. However, it is possible for twig or shoot infection to occur in orchards where little or no blossom infection was found. In these situations the disease pressure is usually low and the damage is more easily controlled. Bacteria can enter the host through both wounds and natural openings such as lenticels. Hail storms often result in severe fire blight outbreaks if inoculum is present in an orchard. Wind damaged leaves are also susceptible to infection.

Fire Blight Management-Cultural Control

Dormant Season: Overwintering cankers should be cut out during the dormant season to reduce sources of bacteria for the next season. Fire blight cankers have either smooth or cracked margins. Both types of cankers should be removed. The smooth margined cankers are harder to see, but they are also more likely to be active than rough margined cankers. Active cankers may enlarge in the spring causing further structural damage. They also provide inoculum for new infections.

One or more separate operations to prune out cankers are recommended. Since cankers may be hard to locate, it is always best to go over the orchard several times. Cankers are most visible on bright, sunny days. Make cuts 15-30 cm below the canker margins. It is not considered necessary to sterilize pruning tools during the dormant season. Do, however, disinfect your tools if spring pruning is extended into late spring when temperatures have warmed up and/or the budburst stage has arrived. These conditions may also reactivate the infectious bacteria in cankers. In fact, an additional inspection for cankers around the budburst stage may reveal cankers that were missed earlier.

Growing Season: Remove current season infections as soon as they are noticed. Prune out infected branches at least 30-40 cm below the visibly diseased part. This is necessary as bacteria are usually present beyond the discoloured area. Dip tools in a disinfectant between each cut. Flag trees that have been pruned, and watch for further symptoms or the development of cankers. Prunings should be removed and burned immediately.

Scout for new fire blight strikes every 3 or 4 days. Frequent scouting will aid removal of new infections before they have a chance to invade the structural wood. Where infections occur on shoots attached to scaffold limbs or the trunk, it is not always possible to cut back 30 cm without sacrificing the limb or even the tree. An option on large trees is to scrape out discoloured inner bark using a hatchet or knife, down to clean wood, and disinfect the cut surface.

Cankers often form at the sites of pruning wounds, where blight was cut out during the summer. Such cankers may be hard to detect. To overcome this problem, it is sometimes recommended that a short (10 cm) stub be left beyond the next healthy spur or branch union when pruning out strikes. Remove the stub later, during dormant pruning. Marking the stubs with bright paint will make them more visible. During severe epidemics, give priority to young trees and high density plantings. Concentrate on salvaging as much of the tree structure and bearing surface as possible. Excessive pruning during the summer will encourage a late flush of growth, which will be susceptible to continued infections.

Summer pruning (other than removal of strikes) should be avoided during a serious outbreak, due to the danger of spreading the disease. If there is any fire blight in the area, disinfect your tools while summer pruning. Avoid pruning during wet weather or when storms are expected within the next 24 hours.

Root suckers and rootstock sprouts may put the entire tree at risk if they become infected. Common dwarfing rootstocks such as M9 and

M26 are highly susceptible to blight. Do not cut rootsuckers or rootstock sprouts during a blight outbreak, because the wounds may become infected. They may be safely removed during the dormant season.

Disinfectants: Good disinfectants for tools include household bleach (eg. Javex, Chlorox), Lysol Concentrated Disinfectant, and PineSol, as well as commercial disinfectants such as Chemprocide. Mix according to label instructions. Bleach can be diluted up to 1:5 with water, and needs to be mixed fresh every day. Tools can either be dipped into, or sprayed with the disinfectant solution. If you use bleach, be aware that it will corrode metal tools and damage your clothing. Ideally, tools should be disinfected after every cut. Dilute disinfectant can also be sprayed on the bark after cutting out an infected branch.

Management Practices: There are several management practices that can reduce the severity of fire blight. Practices that reduce the vigour of the trees tend to reduce fire blight, because young, fast growing tissue is the most susceptible to infection. Don't over fertilize the trees. Excess nitrogen causes vigorous shoot growth, which is more susceptible. Nutrient application should be balanced, preferably based on soil and leaf analysis. Application of fertilizer should also be timed to avoid a late flush of growth, because late season infections are more likely to produce cankers that allow the bacteria to overwinter.

Do not run overhead sprinklers while blossom is present on the tree and weather is favourable for fire blight infection. Overhead irrigation may increase fire blight by splashing bacteria around from tree to tree, and by increasing moisture and humidity levels in the canopy. Cutting back on irrigation may also help to slow down over-vigorous trees. Control insects with sucking mouthparts such as aphids, leafhoppers and pear psylla. These insects can spread fire blight.

Chemical Control: Products registered for fire blight control include the antibiotic streptomycin, fixed copper compounds (copper oxychloride), and copper sulfate (Bordeaux), as well as the biopesticides BlightBan and Bloomtime. These are protectants and therefore must be applied before infection occurs. They will not cure diseased tissue. Always follow label directions and precautions when using pesticides. For comments on spray timing, refer to sections on blossom protection and forecasting fire blight.

Streptomycin 17WP is registered for control of fire blight on both apples and pears. It is effective for blossom blight control during warm (over 18°C) temperatures. Streptomycin provides better control than copper, and does not cause fruit russet. It only provides 2 to 3 days of protection because it breaks down quickly when exposed to sunlight. Streptomycin is most effective when applied as a dilute spray to open blossoms. Better absorption is obtained during slow drying conditions. For best results, do not tank mix streptomycin with other pesticides. Streptomycin has a 50 day pre-harvest interval on apple, and 30 days on pear. Streptomycin should always be stored in a refrigerator, or effectiveness will decline rapidly. Refrigerated product will keep well for two years, with a gradual decline in effectiveness from two to five years.

Fire blight resistance to streptomycin is known to occur in the Okanagan Valley of British Columbia, and throughout most of the Pacific Northwest of the United States. To help preserve any remaining effectiveness, never use streptomycin after the blossom stage, and do not use more than 3 times per season. Alternate with Serenade, Bloomtime or Blightban for resistance management. Never spray streptomycin after an outbreak has already produced shoot blight. Resistance is most likely to occur with repeated use and when pathogen populations are high. Copper oxychloride 50% WP (Clean Crop Copper Spray, Guardsman copper oxychloride) is registered for blossom blight control on pears, crabapple and other susceptible ornamental trees. It is not registered for blossom protection on apples because it will russet the fruit, and also should not be used on Anjou pears for the same reason. Copper oxychloride should not be used on wet foliage. Bartletts sprayed with copper during wet weather have suffered russeting.

Clean Crop Copper Spray is also registered for dormant (silvertip) or post harvest application to apple. Dormant copper applications have been shown to reduce or delay the production of inoculum in overwintering cankers. The addition of a light (summer) emulsifiable spray oil to the dormant copper spray (1 L per 100 L solution) has been shown to increase the effectiveness of the treatment. Bordeaux mixture can also be used for fire blight. It is registered for application to the blossom stage of apple and pear, but should be used with caution due to the potential for russeting. Follow the same precautions as listed for fixed copper sprays. Bordeaux mixture must be prepared on the farm by mixing copper sulfate, hydrated lime and water. The addition of lime to the copper sulfate helps to reduce but may not entirely eliminate the phytotoxicity (burning action) that would occur if copper sulfate was used alone.

To prepare Bordeaux mixture, dissolve 1 kg of tri-basic copper sulfate (Clean Crop Copper 53W), or 2 kg of 25.2% copper sulphate in 1000 L water in the spray tank. Turn on the mechanical agitator. Premix 6 kg of hydrated spray lime in a pail in enough water to make a slurry. Pour the lime slurry through a 0.3 mm screen into the spray tank. Allow 15 minutes of mixing before starting to spray, and keep agitator running while spraying. Bordeaux mixture is not compatible with other pesticides. Apply as a dilute spray only, using 3000 L/ha. Do not use as a concentrate. Do not use Streptomycin after using bordeaux mixture, as the high pH of the Bordeaux will break down the streptomycin.

Note: not all brands of copper sulfate are registered as pesticides. Check the label — it should have instructions on mixing and application for fire blight control.

Caution: Bordeaux mixture is corrosive. Sprayers should be washed thoroughly after use. A protective coat of oil will make clean-up easier by preventing Bordeaux from sticking to paint.

Apogee (prohexadione ca) is a plant growth regulator that is registered for suppression of the shoot blight phase of fire blight on apple. Apogee does not control the blossom blight phase of fire blight, and it is not a bactericide. Apogee will suppress the shoot-blight phase of fire blight when applied 7 to 10 days prior to the onset of conditions favourable for shoot-blight development. For best results, apply when shoot growth reaches 2.5 to 7.2 cm according to label directions. Depending on the cultivar and weather conditions, this may occur during bloom to petal fall. Make subsequent applications at 14-21 day intervals, up to a maximum of 4 applications per season. Apogee is a strong growth regulator and will limit shoot extension. When using the high rate for fire blight suppression, shoot extension may be reduced more than is desired in non-vigorous plantings. Apogee is not registered for use on pear.

Biological Control: Serenade Max, Bloomtime Biological FD and BlightBan C9-1 are three new biopesticides registered for suppression of fire blight on apple and pear. These products are blossom protectants, used during bloom to reduce the incidence of blossom blight. The products consist of beneficial bacteria (*Bacillus subtilis* in Serenade, and *Pantoea agglomerans* in Bloomtime and Blightban) which colonize blossoms and help to prevent fire blight bacteria from getting established. For best results, apply during bloom at the beginning of a warming trend, and follow with a streptomycin spray 2 or 3 days later if warm temperatures continue to favour blossom infection. These biopesticides should be used as part of an integrated fire blight suppression program, which includes using a risk assessment model, as well as cultural controls and streptomycin when necessary. Do not use copper just prior to or during bloom, as it is toxic to the biopesticides and can affect their performance. Bloomtime and Blightban should be kept frozen prior to use.

Serenade Max: Begin applications at early (1-5%) bloom if weather conditions are warm enough to be favourable for fire blight. Repeat at 4-7 day intervals during high risk periods during bloom. Best used in rotation with streptomycin for improved control and resistance management. Also registered for suppression of powdery mildew on apple.

Bloomtime Biological FD Biopesticide: Apply at 15-20% bloom, followed by a second application at full bloom to petal fall. Ensure thorough coverage of the blooms. Do not apply after fruit set. Do not exceed two applications per season.

BlightBan C9-1: Apply at 15-20% bloom followed by a second application at first petal fall to full bloom, and a third application at "rattail" bloom for pear, or post petal fall for apples. Ensure thorough coverage. Use the higher rate under high disease pressure. Maximum of three applications per crop season.

Blossom protection: Temperatures of 18°C and over, along with rain, dew or high humidity, favour blossom infections. In British Columbia, such conditions rarely occur during main bloom in May. They are more likely to occur in June and the summer months with secondary bloom. Spraying for blight is not necessary until daytime temperatures reach 18°C. Spraying at cooler temperatures is a waste of money. Blossom infection may be reduced by spraying with Serenade, BlightBan, Bloomtime, streptomycin or fixed copper. Repeated sprays at 72 hour intervals may be necessary if weather conditions remain warm and moist while blossoms remain on the tree. Do not apply copper to Anjou pears or apples after green tip, due to the danger of fruit russeting. Russeting may also occur on Bartletts if they are wet when sprayed with fixed copper. Copper is not compatible with BlightBan or Bloomtime.

Post-Bloom: Streptomycin is most effective as a blossom protectant spray. It is of limited use once the blossom phase has been completed, and there is little benefit to continued applications after bloom. However, streptomycin or fixed copper applications after petal fall may be helpful when rat-tail bloom is heavy, active cankers are present, and weather is warm and moist. Observe pre-harvest intervals. Note that "trauma blight" (application following hail or wind storm damage) has been removed from the Streptomycin label.

If secondary blossoms could be removed, the risk of fire blight infection would be greatly reduced. It may be feasible to pinch off late bloom on small trees.

Post-harvest: Post-harvest application of fixed copper or Bordeaux mixture in the fall could be considered in orchards with a late flush of growth and an active fire blight problem.

Dormant season: Dormant application of fixed copper or Bordeaux mixture in the spring will help to delay production of inoculum on overwintering cankers. Apply to runoff around the silver tip to green tip stage.

Forecasting Fire blight: There are several models available which can be used to forecast fire blight blossom infection based on the weather. They can be used to determine whether or not sprays are needed to protect bloom. The Cougarblight model, developed by Tim Smith at Washington State University, is very easy to use. It is necessary to note the daily maximum and minimum temperatures while open blossom is present, and determine whether blossoms have been wetted by rain, irrigation or heavy dew. Daily degree-hour estimates are read off a chart based on maximum and minimum temperatures, and added up over a 4-day period. Once a certain threshold is reached, there is risk of blossom infection.

Varietal Resistance

Factors affecting susceptibility: Susceptibility to fire blight can be influenced by many factors, including tree nutrition, irrigation management, soil factors, and cultural practices in the orchard. Young trees are more likely to be severely damaged or killed than older trees. Varieties which have a late or prolonged blossom period are more likely to have bloom on the tree when the weather warms up, and are therefore more likely to suffer from blossom blight. Even relatively resistant trees, such as Red Delicious, may get fire blight after a hail storm or if they are in a mixed planting with susceptible varieties or pears.

Fire Blight Prevention

If you are growing susceptible varieties, be aware that fire blight may strike. Keep your eyes open for symptoms while working in the orchard, and periodically inspect your crop for blight. Take immediate action if you find anything suspicious. Have the disease confirmed by a field advisor or plant pathologist. If fire blight was in your orchard or a nearby orchard the previous year, be sure to monitor for fire blight favourable weather during bloom and apply preventative sprays if necessary. You can minimize the chances of introducing fire blight into your orchard by selecting nursery stock from a reliable source. Ask your supplier if they have had fire blight problems in the nursery. The varieties and rootstocks you grow will also influence the potential for fire blight.

Summary: Fire blight poses a serious threat to new and established orchards of apple and pear. Fire blight has a tendency to appear in unexpected places. It could be well worth your time to regularly scout for symptoms in your orchard, even if you've never had fire blight before. Successful fire blight control requires good management. The following summary lists the major management practices used to control fire blight:

Dormant Season

- remove cankers and dead branches
- delayed dormant copper spray

Blossom Stage

- monitor temperature during blossom
- spray Serenade, Bloomtime or BlightBan biopesticides to reduce blossom infection if blight-favourable conditions are forecast
- spray streptomycin during blossom at 3 day intervals or following biopesticide application only if fire blight risk is high and weather is favourable for blight
- consider Apogee for shoot blight suppression in blocks at risk of fire blight. Must be applied early. Will reduce shoot growth due to growth regulator effect.

Post Bloom

- scout for and remove strikes at regular intervals
- disinfect tools between cuts
- control aphids, leafhoppers and pear psylla

Post Harvest

- apply copper sprays if fire blight is active
- continue to remove new infections

Little Cherry Disease in British Columbia

We now know that there are at least two viruses causing Little Cherry Disease in B.C. Little cherry virus 2 was first detected in British

Columbia in the Kootenays in the 1930s and in the Okanagan in 1969. Little cherry virus 1 was first detected in BC in 1999, but has probably been present for much longer. Little cherry virus 2 is spread from tree to tree by the apple mealybug. Crops on affected trees are not fit for the fresh fruit market because they are lacking in flavour, sweetness, size and colour. A regulation under the B.C. Plant Protection Act provides a mechanism for removal of infected trees.

Distribution: Little cherry virus 2 is present primarily in the Penticton and Creston areas. A few positive trees have also been detected in Summerland and Kelowna in recent years. Little cherry virus 1 appears to be more widely distributed, with positive trees detected from all areas of the Okanagan/Similkameen and Creston valleys.

History of Little Cherry in B.C.

Little cherry disease was first identified in British Columbia in 1933 at Willow Point in the West Kootenay's. The initial infection was most likely caused by infected ornamental flowering cherries which had been planted nearby. Aided by high populations of the apple mealy bug, the disease spread quickly throughout the area, along the Arrow and Kootenay Lakes to Creston and Johnsons Landing in the east and Castlegar and Nakusp in the west. The disease was devastating to the Kootenay cherry industry. The virus spread to virtually all orchards, and infected almost all of the trees within those orchards. Production dropped from 680,000 kg in 1947 to 68,000 kg in 1979, by which time most fruit did not meet packinghouse standards. The Little Cherry Control Regulation was introduced in 1947 under the authority of the B.C. Plant Protection Act to protect the Okanagan cherry industry from the Kootenay little cherry virus.

In 1938, Agriculture Canada introduced a biocontrol agent to the Kootenays to help manage the high population of apple mealy bugs. By the mid 1940's the parasitic wasp had dramatically reduced the mealy bug population. This contributed greatly to reducing the rate of spread of the little cherry virus.

Little cherry disease was first found in the Okanagan in 1969 with the discovery of 49 infected trees in a Penticton orchard. The disease had been introduced to many orchards in the Okanagan by infected cherry rootstock. A control program to identify and remove diseased trees was initiated by the B.C. Ministry of Agriculture and Fisheries in 1970. The annual survey and subsequent tree removal has been effective in reducing the number of diseased trees in the

Okanagan to under 50 per year, from over 300 per year in the late 1970's. Areas where little cherry disease has been found in recent years include Penticton, Naramata, Summerland, Kelowna, Westbank and Oyama. A survey and eradication program for the Creston area was intensified in 1982 by Agriculture Canada in cooperation with the local Noxious Pest Control Committee. The Regional District of Central Kootenay enacted a local bylaw in 1983 to enforce removal of diseased trees. By the 1990's most known infected trees had been removed, and there was a renewed interest in cherry production. In 1994, on the request of Creston area cherry producers, the provincial Little Cherry Control Regulation was amended to include the Creston Valley in the Little Cherry Control Area.

We now know that there are at least two viruses causing Little Cherry Disease in B.C. In recent years natural spread of Little Cherry virus 2 in the Okanagan has been slow due to very low populations of mealybug. Most spread to new locations has been through propagation. It is suspected that some spread within orchards has been due to root grafting. Little cherry Virus 1 appears to be more prevalent, but little is known about how it spreads, other than with propagation.

Symptoms

Cherry trees infected with the little cherry virus 2 produce fruit that is smaller than normal, with poor colour and poor flavour. Fruit symptoms are most pronounced in the Lambert variety; fruit is half to two-thirds normal size, dull red, and pointed, usually with three flat sides tapering toward the blossom end. Fruit fails to develop normal sugar and acid levels, resulting in tasteless fruit lacking in sweetness and flavour. It is common for some fruits on a branch to be more severely affected than others.

Symptoms are similar on other varieties, but less severe and more variable. On varieties with firm, broad-shouldered fruits, such as Van, there is reduction in size and sweetness, but the colour and shape are not usually affected. Infected Bing fruits are similar in shape to Lambert, but there is a tendency to regain almost normal size and fair colour several years after infection. Normal flavour is never recovered. Refer to table for descriptions of symptoms by variety.

In general, symptoms are most severe in the season following the first visible infection. Occasionally infected trees of some varieties will start to produce more "normal" looking fruit again after several years. However, the fruit remains tasteless and insipid, and the virus remains active. Such trees become hard-to-detect carriers of the disease, which continues to spread to other trees. Reddish leaf discoloration occurs on some varieties in late summer or early fall. The variety Sam and Mazzard F12/1 rootstock show a definite leaf reddening in late August or early September. The varieties Bing and Van may display a milder reddish leaf discolouration. Leaf reddening is not a reliable indicator of little cherry disease in orchards because other factors may also cause red leaves.

Little cherry virus may cause infected trees to be less vigourous, especially when trees have been infected for several years, however the general health of the tree usually appears normal. Fruit symptoms are the main manifestation of the disease. Symptoms of little cherry virus 1 appear to be less severe than for Little Cherry Virus 2 in BC orchards, but this is not yet well documented.

Other disorders that may resemble little cherry;

- There are several other disorders and conditions that may cause cherry fruit to be smaller than normal. When scouting for symptoms of little cherry virus, it is best to look for symptoms on healthy limbs with a normal crop load. It is very difficult to identify little cherry infections on neglected or winter damaged trees.
- Winter damage may cause weak branches with sparse foliage and small fruit.
- Zinc deficiency may cause both leaves and fruit to be small and discoloured.
- A heavy fruit crop will often result in cherries that are smaller than normal. Flavour should not be affected.
- Western X disease also causes small fruit with poor flavour. This disease does not spread as fast, often remaining in only some of the limbs of a tree for many years. Fruit eventually becomes almost whitish in colour, and has a bitter taste. Fruit infected by little cherry is tasteless, but not bitter.

Life Cycle

Little cherry virus 2 is spread from tree to tree by the apple mealybug. There are no other known insect vectors. The virus is also readily transmitted by all types of grafting. Transmission by pollen, seed, in the soil or by pruning tools has not been demonstrated.

Little cherry virus 1 can also be spread by propagation. The insect vector is not known.

Ornamental flowering cherries are symptomless carriers of the disease, and are prohibited in the Okanagan and Creston valleys.

Mealybug life cycle: Apple mealy bug (*Phenacoccus aceris*) is the only known vector of little cherry virus. Other than spreading the virus, low populations do not normally cause any direct damage to cherry trees. Hosts include a wide range of ornamental and fruit trees including apple, pear, plum, cherry, currants, maple, birch, willow and white ash.

The apple mealybug overwinters as partially grown nymphs in small, white cocoons (approximately 2 mm long) which are hidden in bark crevices. Females emerge during the first warm weather in the spring, and begin feeding on the undersides of twigs. Winged males emerge in May to seek mates. Eggs are laid in sheltered areas during June and July, and newly hatched nymphs (or crawlers) appear from July to August. Crawlers feed on the undersides of leaves and excrete honeydew. The major period of mealybug movement occurs during the summer when young nymphs are feeding on the leaves and can be transported by wind. As they mature, nymphs begin to feed on fruit and the bark of small twigs and spurs. Hibernation begins in October. There is one generation per year.

Adults are yellow to orange in colour, about 3 mm long, and covered with a granular, white, waxy material. Eggs are yellow and found inside masses of white, cottony wax. Nymphs are oval, flattened and yellow with red eyes. They can be confused with the crawler stage of scale insects but the latter do not have red eyes. Mature nymphs are about 2.8 mm long.

Control

Little Cherry Disease can be controlled by following the guidelines below. Look for symptoms at picking time. If you suspect little cherry, please contact your field advisor, or the BCMAL plant pathologist in Kelowna.

1. Avoid planting infected stocks or using budwood from trees that have not been tested for freedom from viruses. Select only nursery stock that has been produced from certified virus-free material. Material labelled "best available" has not been completely tested for viruses, and may introduce Little Cherry or other serious virus diseases into your orchard.
2. Remove all trees known or suspected to be infected with little cherry disease to limit spread within your orchard. Carefully

inspect fruit quality on trees in high risk areas for small fruit symptoms. Remove suspect trees promptly after cherry harvest before the active mealy bug crawlers have a chance to further spread the disease. Stumps should be prevented from sprouting, and removed before spring. Trees with little cherry virus must be removed under the authority of the B.C. Plant Protection Act.

3. Do not interplant cherry trees amongst older cherry trees that may be carrying the virus (symptoms are not always obvious). It is best to remove whole blocks of older cherry trees before replanting to cherry, particularly in orchards or districts with a history of little cherry virus. Note: this will also help to delay introduction of other cherry virus diseases.
4. Do not grow or import ornamental Japanese flowering cherries, as these trees are often symptomless carriers of the little cherry virus. The Little Cherry Control Regulation prohibits the sale or growth of flowering cherries in the Okanagan, Similkameen and Creston valleys.
5. Eliminate cherry seedlings in and near the orchard, as they can be symptomless carriers. Wild *Prunus emarginata* (bitter cherry), which is common in the Creston area, may also harbour the little cherry virus.
6. Control apple mealybug. All cherry growers with orchards where little cherry virus 2 has been found should apply sprays at all four recommended stages listed below. Growers in areas with a high incidence of little cherry (Penticton, Creston) should also follow the program to protect their trees.

Mealybug Control Program

The following sprays are recommended to control the apple mealybug in cherry orchards where little cherry disease has been found. Also refer to the Tree Fruit Production Guide and the product labels for information on application rates and safety precautions for using the insecticides listed below.

1. *Dormant:* Apply oil plus diazinon. This spray controls overwintering nymphs. Use high volume application techniques as recommended for San Jose scale control.
2. *Petal-Fall:* Admire, Diazinon, Guthion, or Sniper used at this stage are effective against mealybugs. High volume airblast or hand gun applications are the most effective.

3. *Summer:* The major period of mealybug movement from tree to tree is during the summer when nymphs are on leaves and can be transported by wind. Nymphs are also the easiest to kill with chemicals. Low volume air blast sprays of Admire, Cygon, Lagon, Guthion, Sniper or Diazinon applied for control of cherry fruit fly are also effective for the control of apple mealybug nymphs. The closer to harvest these materials are applied, taking the minimum days before harvest into account, the more effective is the mealybug control.
4. *Post Harvest:* A post harvest spray of diazinon is the most effective control of nymphs on leaves, but Admire and Cygon or Lagon applied for cherry fruit fly will also provide adequate control.

Little Cherry Control Regulation: The Little Cherry Control Regulation of the B.C. Plant Protection Act regulates the destruction of trees with little cherry virus, the cultivation of ornamental flowering cherries and the transportation of cherry stock.

Chapter

11

Pest Management Strategic Plan for Pulse Crops

Background

Edible seeds of legumes are known as pulses. They include peas, beans, lentils, and chickpeas. The term "pulse" comes from the Latin word "puls," which means a thick soup. For the purposes of this document, the term "pulse crops" will represent chickpeas, lentils, and dry peas. No other pulse crops will be included in the discussions. All three of these pulse crop species are a part of the larger plant family known as the Fabaceae or legume family. The Fabaceae family includes about 600 genera and 13,000 species, making it the third largest family within the plant kingdom. Further taxonomic classification puts the pulse crop species in the Faboideae subfamily and the Fabeae tribe. This family includes cultivated species such as alfalfa, soybeans, and many edible beans. Pulses are considered environmentally friendly because of their reduced dependence on fossil fuels. Instead of requiring fertilizer applications, they are able to obtain much of their nitrogen requirement from the atmosphere by forming a symbiotic relationship with Rhizobium bacteria in the soil. Pulse crops' low crop residues and low carbon-to-nitrogen ratios eliminate the need for burning and make rotating to the next crop using reduced tillage very easy.

Chickpeas (Cicer Arietinum)

Chickpeas are an annual grain legume and are believed to be one of the first legumes cultivated by humans. Among pulse crops marketed as human food, world chickpea consumption is second only to dry beans. They are classified as two types, "desi" or "kabuli," based on seed size, colour, and the thickness and shape of the seed coat. About 90 percent of chickpeas, the majority of which are desi types, are consumed in India. Desi types are usually smaller, angular seeds with thick seed coats that range in colour from light tan and speckled to solid black. Desi chickpeas require a specialized seed-coatremoval process if used for human food. The process, called decortication, requires adjusting the moisture level of the seeds to facilitate the mechanical removal of the thick seed coat. The seeds, which after decortication resemble a small yellow pea, are processed into numerous South Asian ethnic food products.

Kabuli types, also known as "garbanzo beans" in the United States, are larger seeds with paper-thin seed coats that range in colour from white to pale cream-colour to tan. In North America, most kabuli chickpeas are marketed as canned chickpeas for salads. Kabuli chickpeas are also marketed as dry chickpeas and ground flour for baking purposes. In the United States, chickpea production is located in Idaho, Washington, Oregon, Montana, North Dakota, South Dakota, Nebraska, Colorado, and California. Production has expanded to over one million acres in the Canadian Prairies (primarily Saskatchewan).

Chickpeas are a good livestock feed. Feed values and feeding studies for kabuli and desi chickpeas have been compared to peas, barley, grain, and soybeans. Chickpeas have higher oil content than other pulse crops and provide slightly more protein and slightly less energy than peas.

Chickpea plants are erect with primary, secondary and tertiary branching, resembling a small bush. They flower profusely and have an indeterminate growth habit, continuing to flower and set pods as long as conditions are favourable. Pod set occurs on the primary and secondary branches and on the main stem. The individual round pods generally contain one seed in kabuli types and often two seeds in desi types. Chickpeas have deeper taproots than peas and lentils. This has an impact in moisture-deficient areas.

Lentils (Lens Culinaris)

In the United States, lentils are grown primarily in the Palouse (Region 1) and upper Midwest (Regions 3 and 4). There are five to six market classes of lentils grown in these regions, based on seed size, cotyledon colour, and seed coat coloration. In Canada, lentils are grown primarily in Saskatchewan. This province is the world's largest exporter of lentils, with 1.7 million acres in production. Smaller acreages are grown in Alberta and in Manitoba. Canada also produces five to six market classes of lentils. Lentil plants are herbaceous, with slender stems and branches. Plant height ranges from 12 to 15 inches for most varieties, but can vary from 8 to 30 inches depending on variety and environment. Plants have a slender taproot with fibrous lateral roots. Rooting patterns range from a many-branched, shallow root system to types that are less branched and more deeply rooted. The taproot and lateral roots in surface layers of the soil have nodules that vary in shape from round to elongate. Stems of lentil plants are square and ribbed and usually thin and weak. Branches arise directly from the main stem and may emerge from the cotyledonary node below ground or from nodes above ground. Leaves are relatively small compared to those of other large-seeded food legumes. Pods are oblong, laterally compressed, 6-20 mm long and 3-10 mm wide, and usually contain one or two lens-shaped seeds. Seed diameter ranges from 2 to 9 mm and colors range from light green or greenish red to gray, tan, brown or black. Purple and black mottling and speckling of seeds are common in some varieties. Varieties grown in North America include large green types with seed diameters from 6 to 7 mm and with yellow cotyledons (Laird, Brewer, Richlea, Mason, Pennell, Merrit, and others), small red cotyledon types (Crimson, Robin), Spanish browns (Pardina), small yellow types (Eston), and large red types (Redchief). Other types may be grown to a lesser extent. The lentil crop in North America is planted in early spring and harvested in late summer.

Dry Peas (Pisum Sativum *and* P. Sativum *ssp.* Arvense)

U.S. dry pea production is located in Washington, Idaho, Oregon, Montana, and North Dakota. Dry peas or "field peas" are a cool-season annual crop. They differ from succulent peas in that dry peas are marketed as dry, shelled products for either human food or livestock feed, whereas succulent peas are marketed as fresh or canned vegetables. There are two main types of dry peas. One type has normal leaves and vine length of 3 to 6 feet. The second type is semi-leafless (actually with modified leaflets) with shorter vine lengths of 2 to 4 feet.

In Canada, dry peas are grown in all three prairie provinces (Alberta, Manitoba, and Saskatchewan) on 3.5 million acres. Dry pea has been the leading alternative crop as farmers move to diversify crop production in the prairies. Mainly yellow and green peas along with other minor classes are grown, with approximately two-thirds of production in yellow types. Most varieties have a white flower, with recent years seeing an increase in the production of northern-European origin varieties with the semi-leafless type. Dry peas emerge and perform well in a variety of seedbeds including direct seeding into grain residue. Dry peas are typically grown following cereal crops. As with the other pulse crops discussed here, most dry peas are spring-seeded; optimal planting dates range from mid-March to mid-May when soil temperatures are above 40 degrees F (4 degrees C). In most years, delayed planting lowers quality and seed yield. Dry peas are adapted to grow during the cool season when evapo-transpiration is minimal. In Region 1, they rely on stored soil moisture for a large part of their growth cycle. Fall-seeded peas (Austrian winter peas) are grown in some parts of the United States and there is potential for development of edible types of dry peas and lentils to be fall-sown in the United States as well.

Depending on variety, dry peas start flowering after a specific number of nodes are reached and flowering continues until drought or nitrogen deficiency brings it to an end. Dry pea varieties have either a determinate or indeterminate flowering habit. Determinate varieties mature in 80 to 90 days, indeterminate varieties in 90 to 100 days. Dry pea harvest begins in late July when pods are dry and seed moisture is 8% to 18%, depending upon the growing region. They are combined directly in the field. A timely harvest is important to avoid post-maturity disease, seed bleaching, and seed shatter.

EPA and FQPA

The U.S. Environmental Protection Agency (EPA) is now engaged in the process of reregistering pesticides under the requirements of the Food Quality Protection Act (FQPA) of 1996. The Agency is examining dietary, ecological, residential, and occupational risks posed by certain pesticides. EPA's regulatory focus on the organophosphate (OP), carbamate, and B2 carcinogen pesticides has created uncertainty as to these pesticides' future availability to growers. At some point, the EPA may propose to modify or cancel some or all uses of these chemicals on pulses. The regulatory studies that EPA requires place significant burdens of time and expense on registrants. In cases where these burdens outweigh the commercial viability of the product, companies may opt to voluntarily cancel certain registrations for pulse crops. The U.S. Department of Agriculture (USDA), the EPA, the land-grant universities, and the pulse industry need to proactively identify research, regulatory, and educational needs in order to reduce the reliance on certain pesticides and identify effective alternatives.

PMRA

In Canada, pesticide registration is a responsibility of the Pest Management Regulatory Agency (PMRA) of Health Canada. Registration is carried out under the authority of the Pest Control Products Act (PCPA) and involves conducting science-based health, environmental, and value (including efficacy) assessments of each pesticide before deciding if it should be approved for use in Canada.

PMRA is also involved in the re-evaluation of older pesticides, using current scientific approaches to examine the continued acceptability of older active ingredients and their end-use products. This re-evaluation program may have implications for pulse crops. Recently, however, a new initiative to help increase the availability of reduced-risk pesticides, including minor use products, was announced by the Ministers of Health and Agriculture and Agri-Food. This initiative will assist in the registration of pesticide products that are used in such small quantities that manufacturers find the sales potential is not sufficient to seek a registration in Canada.

Harmonization

In the United States and Canada, EPA and PMRA are working together to develop common or compatible international approaches to pesticide review, registration, and standard-setting for pesticides, while maintaining current high levels of protection of public health and the environment and supporting the principles of sustainable pest management. EPA and PMRA believe that making pesticide regulatory programs more consistent internationally will lead to: (1) improved food safety; (2) reduced regulatory burden on national governments; (3) strengthened scientific procedures; (4) fewer trade problems; and (5) better access to lower-risk pesticides and integrated pest management (IPM) programs that help reduce reliance on conventional pesticides.

To North American farmers, harmonized regulatory systems mean greater efficiencies on the farm through faster access to IPM crop protection tools.

The Work Group and the PMSP

A work group consisting of U.S. and Canadian growers, commodity groups, pest control advisors, regulators, and university specialists along with representatives from USDA, EPA, AAFC, and PMRA met for two days in Saskatoon, Canada in June of 2002. The purpose of the meeting was to identify the needs of pulse growers in the two countries with reference to possible regulatory actions regarding pesticides. The outcome of this exercise resulted in a list of critical needs, general conclusions, and tables listing the timing of operations and the efficacies of various management tools for specific pests. These materials have been compiled and reviewed by task force members and are presented in this Pest Management Strategic Plan (PMSP).

This document and its appendices are intended to serve as a comprehensive foundation for pest management transition in pulse crops in the United States and Canada.

The following list summarizes those needs determined by the task force to be the most critical to U.S. and Canadian pulse crop pest management. Overall top priorities are listed first, followed by the top priorities for each of the three pulse crops under examination. The remainder of the document is organized by crop production period from pre-plant through post-harvest. Each crop production period is subdivided by pest (e.g., weed, disease, insect) with critical needs in the areas of research, regulation, and education identified at the end of each section.

U.S. Overall Top Priorities

Research

- Ascochyta management and control
- Biology/ecology of winter pulses
- Expansion of post-emergent herbicide labels
- Fungicides for foliar disease control
- Insecticide research, particularly dimethoate alternatives.

Regulatory

- Retain dimethoate registration
- Harmonization of products, cropping zones, maximum residue levels (MRLs), etc.
- Full label for sulfentrazone (Spartan), 2,4-DB, thiabendazole (LSP, Mertect), and pyridate (Tough); note that pyraclostrobin (Headline) and azoxystrobin (Quadris) were originally on this list but their registrants announced the products' registrations in March and April 2003, respectively
- Get pulse crops on EPA's PR 97-2 list for all relevant crop/pest combinations.

Education

- Put herbicide mode of action on label
- Tell EPA and registrants what pulse industry needs
- Educate growers on available resources
- Transfer predictive models for ascochyta blight to growers
- Update PMSP in 2 years.

U.S. Crop-specific Priorities

Chickpeas

Research:

- Develop post-emergent broadleaf herbicides
- Develop larger kabuli-type chickpea varieties
- Develop earlier-maturing varieties
- Develop cost-effective fungicides
- Investigate biology and epidemiology of ascochyta blight for modeling risk management
- Increase breeding efforts (blight resistance particularly)
- Develop alternative to glyphosate (Roundup Ultra).

Regulatory:

- Work with regulators to decrease delay in Section 18 fungicide approvals, particularly those involved with disease resistance management
- Expedite full registration of fungicides
- Register additional active ingredients on chickpea in general (use crop grouping approach)
- Get pulse crops on EPA's PR 97-2 list for all relevant crop/pest combinations
- Expedite full label registration of sulfentrazone (Spartan) and pyridate (Tough); note that pyraclostrobin (Headline) and azoxystrobin (Quadris) were originally on this list but their registrants announced the products' registrations in March and April 2003, respectively.

Education:

- Attract priority attention from fungicide registrants
- Industry education about registration techniques and funding sources for research

- Conduct disease identification courses
- Transfer predictive models to growers.

Lentils

Research:

- Increase breeding efforts (e.g., disease resistance, shatter resistance, etc.)
- Investigate biology and ecology of winter lentils (crop-pest interaction)
- Develop pre- and post-emergence broadleaf herbicides
- Develop an alternative to dimethoate
- Investigate environmentally friendly/efficacious weed control in postharvest situations (i.e., winter wheat or fallow)
- Develop seed treatment for aphid control
- Develop a forecasting model for pea aphid/lygus bug (virus transmission problems)
- Develop additional herbicides in wheat that will not harm lentils grown in a later rotation.

Regulatory:

- Obtain 2,4-DB registration
- Retain dimethoate registration
- `Need more chemistries (all types) for lentils
- Decrease 3-year requirement between pulse rotations for crop insurance coverage (ND, MT)
- Expedite bifenthrin (Capture 2E) and lambda cyhalothrin (Warrior) registrations
- Harmonization of chemical registrations
- Expedite full registration of thiabendazole (LSP, Mertect).

Education:

- Inform EPA about the critical use of dimethoate
- Provide additional information on weed control to growers, especially to new growing areas
- Get registrants to add lentils to labels when data is available for crop groups
- Get information to crop insurance people
- Inform growers about aphid models
- Develop better communication with registrants.

Dry Peas

Research:

- Develop cost-effective control methods for weevils in field
- Develop seed treatment for aphid and disease control
- Develop pea weevil materials that will not cause flaring of aphid populations
- Increase breeding efforts for pest management
- Develop additional cost efficient fungicides
- Develop new pre- and post-emergent herbicides
- Develop a replacement for dimethoate
- Conduct biological and ecological studies of winter peas
- Investigate vertebrate control (voles).

Regulatory:

- Harmonization of imazethapyr (Pursuit) registration and product pricing
- Expedite sulfentrazone (Spartan) registration
- Need more fungicides registered
- Expedite bifenthrin (Capture 2E) and lambda-cyhalothrin (Warrior) registrations
- Retain dimethoate registration
- Obtain different allowable residue levels when crop used as feed
- Extend seed treatment labels to include peas.

Education:

- Start tracking pesticide use data
- Educate growers about vertebrate control
- Inform EPA about dimethoate criticality
- Provide additional information on weed control to growers, especially to new growing areas
- Get registrants to add peas to labels when data available for crop groups
- Get information to crop insurance people
- Get growers information on aphid models
- Develop better communication with registrants
- Attract priority attention from fungicide registrants
- Educate industry about registration techniques and funding sources for research

- Conduct disease identification courses
- Transfer predictive models to grower.

Canada Overall Top Priorities

- Develop super-crop groupings (combinations of crops with similar pest management needs for joint regulatory consideration)
- Pesticide harmonization
- Additional IPM crop protection tools for disease management
- Additional IPM crop protection tools for broadleaf weed management
- Work with international trade partners so that MRLs don't become trade irritants
- Catch up pesticide registration and on-farm food safety; develop list of products that need to be registered. These will include the following immediate priorities: (1) ethalfluralin (Edge, Sonalan) on chickpeas, (2) deltamethrin (Decis) and permethrin (Ambush) on chickpeas, (3) bentazon (Basagran) applied pre-bloom on peas for sow thistle control, and (4) broadleaf weed control in pulse crops.
- Re-evaluate zone maps for residue trials
- Develop crop profiles
- Define, monitor, and document IPM practices in pulses
  - o Educate regulators
  - o Educate public
  - o Standardize measurements.

Canada Crop-specific Priorities

Chickpeas

Research:

- Enhance ascochyta-resistance breeding programs
- Improved control/management tools and practices for ascochyta:
  - o Modeling/forecasting/monitoring tool
  - o Fungicide chemistry
  - o Fungicide rotations.
- Increase breeding for early maturity
- Develop broadleaf weed control tools
- Improve harvest management tools (desiccants, growth regulators, defoliants)

- Investigate wireworm management through seed treatments or insecticides
- Develop alternative reduced-risk fungicides products and/or methods.

Regulatory:

- Expand labels for ethalfluralin (Edge, Sonalan), diquat, imidazolinones
- Register sulfentrazone (Spartan), strobilurins, and pyridate (Tough)
- Harmonize (with U.S.) and streamline registration process
    - o Compressing trial zones
    - o Increase acceptance of U.S. data
    - o Expedite setting up a Canadian IR-4
- Pulse industry to work with PRMA on regulatory flexibility to maintain pulse crops in the minor use system, to obtain product registration in pulse crops
- Industry to explore potential of picking up minor pesticides
- Improve coordination within Canada to generate data to fill requirements for minor use registrations
- Examine "novel trait" plant definition.

Education:

- Educate growers about disease management, resistance management, disease identification, and field scouting
- Improve stakeholder awareness of regulatory requirements for pulses
- Educate growers about rotations and other cultural pest management strategies
- Educate growers about pesticide persistence
- Develop crop profiles
- Harmonize (with U.S.) and streamline registration process:
    - o Compressing trial zones
    - o Increase acceptance of U.S. data
    - o Expedite setting up a Canadian IR-4.
- Pulse industry to work with PRMA on regulatory flexibility to maintain pulse crops in the minor use system, to obtain product registration in pulse crops

- Industry to explore potential of picking up minor pesticides
- Improve coordination within Canada to generate data to fill requirements for minor use registrations..

Lentils

Research:

- Enhance ascochyta and anthracnose resistance breeding programs
- Continue research in control/management tools and practices for disease:
    - o Modeling/forecasting tool
    - o Fungicide chemistry
    - o Fungicide rotations.
- Explore broadleaf weed control tools
- Evaluate potential of new strobilurins.

Regulatory:

- Label expansion for imazethapyr (Pursuit) and other imidazolinones
- Register strobilurins
- Harmonize (with U.S.) and streamline registration process:
    - o Compressing trial zones
    - o Increase acceptance of U.S. data
    - o Expedite setting up a Canadian IR-4.
- Pulse industry to work with PMRA on regulatory flexibility to maintain pulse crops in the minor use system, to obtain product registration in pulse crops
- Industry to explore potential of picking up minor pesticides
- Improve coordination within Canada to generate data to fill requirements for minor use registrations
- Examine “novel trait” plant definition.

Education:

- Educate on disease management, resistance management, disease identification, and field scouting
- Stakeholder awareness of regulatory requirements for pulses
- Educate on rotations and other cultural pest management strategies

- Educate on pesticide persistence
- Develop crop profiles
- Harmonize (with U.S.) and streamline registration process:
  - o Compressing trial zones
  - o Increase acceptance of U.S. data
  - o Expedite setting up a Canadian IR-4.
- Pulse industry to work with PRMA on regulatory flexibility to maintain pulse crops in the minor use system, to obtain product registration in pulse crops
- Industry to explore potential of picking up minor pesticides
- Improve coordination within Canada to generate data to fill requirements for minor use registrations.

Dry Peas

Research:

- Develop new annual broadleaf products to replace imidazolinones
- Investigate disease management tools: modeling, resistance, decisionmaking tools
- Develop harvest aid management tools:
  - o Shorter grower season areas
  - o New market classes.
- Continue breeding for mycospharella resistance
- Develop new stand-up leafy peas.

Regulatory:

- New labels: strobilurins, broadleaf herbicides
- Harmonize (with U.S.) and streamline registration process:
  - o Compressing trial zones
  - o Increase acceptance of U.S. data
  - o Expedite setting up a Canadian IR-4.
- Pulse industry to work with PRMA on regulatory flexibility to maintain pulse crops in the minor use system, to obtain product registration in pulse crops
- Industry to explore potential of picking up registration duties for minor pesticides
- Improve coordination within Canada to generate data to fill requirements for minor use registrations
- Examine “novel trait” plant definition.

Education

- Educate growers about harvest management
- Educate stakeholders about regulatory requirements for pulses
- Educate growers on rotations and other cultural pest management strategies
- Educate growers on pesticide persistence
- Develop crop profiles
- Educate general public on pest management and role of pesticides in pulse crop production
- Implement professional development requirement for regulators.

Pest Management Strategic Plan Foundation

The remainder of this document is an analysis of pest pressures during the various growth stages of pulse crops. Key control measures and their alternatives (current and potential) are discussed. Differences between production regions throughout the United States and Canada are discussed where appropriate. All three pulse crops are discussed in each section, with differences and similarities in pest control noted.

Pre-Plant

It is important to consider field history in pulse production. The decision to grow pulses in a given field is typically made a year or two in advance to allow for proper site preparation. Therefore, the "pre-plant" period for any field includes all production seasons subsequent to the previous pulse crop and the late fall, winter, and early spring just prior to the planting of the pulse crop.

Many herbicides used in small grain production may persist in the soil, resulting in pulse crop injury and yield loss. Rotational intervals depend in part upon how long herbicides remain in the soil. Factors that affect herbicide persistence include soil pH, moisture, temperature, texture, and organic matter. In areas with a dry climate and short growing season, herbicides generally degrade slower than in warmer, moister areas. Sulfonylurea herbicides (Ally, Amber, Canvas, Finesse, Glean, and Peak) persist longer in higher pH soils. When soil pH exceeds 7.5 to 7.9, sulfonylurea residues may remain in the soil much longer than described on the label. Under such conditions, a field bioassay is required the year before seeding pulses.

The rotation of pulse crops with other crops is an integral part of the pre-plant pest control strategy. Pulses are best grown following a cereal rather than another crop that may harbor pulse diseases. Pulse crops are susceptible to diseases that can overwinter in the soil and in stubble. These considerations are important in management of weeds and diseases and in minimizing residual herbicide injury to the crop. Pulse crops offer several agronomic and economic advantages for the producer. Cereal crop yields often increase when planted after legumes because cereal pest (disease, insect, and weed) cycles have been disrupted. Legume crops enable use of different herbicides than the cereal crops to clean up grassy weeds. The legumes conserve soil moisture and limit soil erosion by offering an option other than summer fallow. Finally, pulses increase the nitrogen content of the soil.

Chickpeas can be seeded into standing or tilled stubble and fallow. Lentils and dry peas emerge and perform well when planted in a variety of seedbeds but most are direct seeded into grain residue. They typically are grown following winter wheat or spring barley. Cereal stubble that is fall plowed or chiseled is cultivated for weed control then harrowed and rolled.

Weeds

Weed control is critical for good production of pulses as well as for the health of the subsequent crops in the rotation. Because they are slow to establish and produce limited vegetative growth, pulse crops are not very effective competitors with weeds. For this reason, growers avoid fields with a history of broadleaf perennial weed problems. Growers practice crop rotation, which may include chemical fallow, to decrease weed populations. Cultural practices to form a smooth, firm seedbed are used to reduce winter annual weeds. But seedbed preparation alone is not sufficient for management of annual weeds.

Control of annual weeds usually begins in the fall or spring prior to planting with nonselective herbicide applications or tillage operations followed by herbicide applications. Perennial weed control begins 1-3 years earlier with herbicide applications or tillage operations. Fall or spring pre-seeding tillage is used to control winter annual and biennial weeds and to activate some soil-incorporated herbicides. Spring tillage controls earlyemerging, summer annual weeds. Shallow tillage avoids bringing weed seeds up near the soil surface where they are likely to germinate. Excessive tillage dries the seedbed making shallow seeding less effective; it also leads to soil erosion.

Weeds are also managed with stale seedbed techniques, such as delaying seeding, allowing weeds to emerge, and then destroying them with either tillage or a non-selective herbicide. These techniques are not foolproof, because weeds emerge throughout the growing season and warm-season annual weeds may be favored by delayed seeding. In addition, delayed seeding will diminish yield potential and quality. Sowing pulses into clean fields is preferred, but pulses are frequently seeded on stubble, where potential weed competition is often high and can be complicated by volunteer plant growth. If pulse crops are directly seeded into stubble with reduced-tillage methods, the previous crop is very important. Volunteer canola or tame mustard is especially important to control as these can smother a pulse crop. Volunteer cereal control is important, as the seeds from any plants that escape can be difficult to separate from lentils.

The move to direct seeding has reduced overall pesticide load, but pulse production still requires pesticide use to maintain yields. Direct seeding practices have reduced the weed seed bank, which in the long run may reduce the total herbicide load. Few herbicides are registered for managing weeds in pulse crops, especially broadleaf weeds. Perennial broadleaf weeds like field bindweed and Canada thistle are not successfully controlled during the fall before seeding pulses, so problems must be addressed in preceding crops. Annual broadleaf weeds such as kochia and Russian thistle are not successfully controlled in small grains or fallow rotations before seeding pulses. Because of the tumbling nature of these weeds, they are likely to re-infest fields by wind movement.

Grass weeds are easier than broadleaf weeds to manage in pulses; rotating to pulse crops may improve grass weed management in small grains. Besides reducing yields through competition, weeds present other problems in pulse production. Weeds can contribute exudates at harvest that stain pulses, reducing quality. Weeds also interfere with mechanical harvesting.

Cultural Practices

Employment of certain cultural practices helps to control weeds. Use of clean equipment is an essential practice for all pulse growers. Selecting proper fields and being mindful of rotation (i.e., which crops were grown previously, which weeds were predominate, etc.) also helps. Growers find that extending rotations so that the legume crop is grown only once every three to four years helps keep weed populations more manageable. The way in which fertilizer is used (i.e., timing, quantity, etc.) will affect growth of weeds as well as the crop. It is important that the pulse crop establish quickly and at an adequate density to compete successfully with emerging weeds.

Some practices control certain weeds while encouraging others; growers base their decisions on the circumstances specific to their field in a given year. For example, deep tillage can be useful in discouraging some weeds, while low-till or no-till methods discourage wild oat, nightshade, and green foxtail. Removing crop residues through either field burning or silaging will remove much of the seed bank, but is not generally cost-effective.

While cultural practices confer many benefits, employment of cultural practices alone cannot effect complete weed control. Most weeds are managed by the use of integrated weed management, combining cultural practices with the judicious use of herbicides.

Direct-seed systems create heavy reliance on glyphosate, which may result in resistance. Herbicide-resistant crops complicate weed control. The potential release of Roundup-Ready wheat varieties means that control systems other than the current standard of applying glyphosate for pre-plant burndown will have to be developed.

Also, the practice of controlling volunteer canola or cereals with imazethapyr or imazamox will not be effective on Clearfield or imidazolinone-resistant varieties.

Chemical Controls

In this section only, due to the relatively large number of chemical options (not all of which are effective), the variability between regions, and the differences between the three crops, chemical control strategies are presented in bulleted lists sorted by crop, then by region.

Chickpeas

Region 1:

- Glyphosate (Roundup Ultra) is applied pre-plant fall or spring for annual weeds, though it is not very effective on buckwheat, prickly lettuce, or any annual weed not yet emerged.
- Ethalfluralin (Edge, Sonalan) and pendimethalin (Prowl) are registered for buckwheat control, but herbicide carryover onto winter wheat that is planted within 6 months is unacceptable. In effect, nothing is available for buckwheat control.
- Imazethapyr (Pursuit) is most commonly used at the pre-plant stage for annual broadleaf weeds and some grasses but it may cause crop damage. Effective control is lost at reduced rates. Imazethapyr (Pursuit) is weak on wild oat, dog fennel, prickly lettuce, and common lambsquarters.

- Pendimethalin (Prowl), when applied in combination with imazethapyr (Pursuit), broadens the broadleaf control, but does not control wild oat or other grasses.
- Triallate (Far-Go, Avadex BW) is used only for wild oat control and must be incorporated twice.
- Metribuzin (Sencor, Lexone DF) is also used for control of annual broadleaf weeds. Pre-plant applications of metribuzin (Sencor, Lexone DF) do not provide effective control and result in crop damage if applied in areas with heavy moisture (1/2-inch rain within 48 hrs. of application).
- Dimethenamid (Frontier 6EC), which was recently registered for annual grasses and broadleaf, is weak on dog fennel.
- S-metolachlor (Dual Magnum, Dual II Magnum) is not applied often because of plant-back restrictions on winter wheat.
- Trifluralin (Treflan) is labelled, but seldom used due to carryover and the need for incorporation.

Region 2: No chickpeas grown in Region 2.

Region 3: Canada

- Glyphosate (Roundup Ultra) is applied pre-plant fall or spring for annual weeds, though it is not very effective on buckwheat, prickly lettuce, or any annual weed not yet emerged.
- High rates of glyphosate (Roundup Ultra) are applied for foxtail barley and dandelion control.
- No other pre-plant herbicides are presently registered.
- Controlling volunteer Round-up Ready canola is a concern. U.S.
- Imazethapyr (Pursuit) is not used in Montana, due to crop injury potential based on low soil organic matter and low moisture. Imazethapyr (Pursuit) is used in North Dakota.
- Sulfentrazone (Spartan), a key herbicide for kochia, Russian thistle, pigweed, and wild buckwheat, is used in Montana, North Dakota, and South Dakota under a Section 18, but it only provides suppression for wild buckwheat.
- North Dakota does not have registrations for metribuzin (Sencor, Lexone) or dimethenamid (Frontier 6EC).
- DNAs are used for a narrow spectrum of broadleaf weeds and some grassy weeds. They do not sufficiently control mustards, kochia, Russian thistle, or wild buckwheat. Downy brome control is important.

- Ethalfluralin (Edge, Sonalan) and trifluralin (Treflan) labels require two incorporations, therefore these options don't fit into low-till or notill systems.
- Pendimethalin (Prowl) and sulfentrazone (Spartan) must be incorporated with rain (which doesn't always occur timely).

Region 4: Very few chickpeas are grown in Region 4.

Lentils

Region 1:

- Glyphosate (Roundup Ultra) is applied pre-plant fall or spring for annual weeds, though it is not very effective on buckwheat, prickly lettuce, or any annual weed not yet emerged. In effect, nothing is available for buckwheat control.
- Pendimethalin (Prowl) is registered for buckwheat control, but carryover of herbicide is unacceptable because winter wheat is planted within 6 months.
- Imazethapyr (Pursuit) is most commonly used at the pre-plant stage for annual broadleaf weeds and some grasses but may cause crop damage. Reduced rates do not provide effective control. It is weak on wild oat, dog fennel, prickly lettuce, and common lambsquarters.
- Pendimethalin (Prowl) is applied in combination with imazethapyr (Pursuit). It broadens the broadleaf control but does not control wild oat and other grasses.
- Triallate (Far-Go, Avadex BW), which uses a different mode of action, is used only for wild oat control, but controls no other species and must be incorporated twice.
- Metribuzin (Sencor, Lexone DF) is also used on annual broadleaf weeds. Pre-plant applications are not very effective and result in crop damage with heavy moisture (1/2 inch rain within 48 hrs. of application).
- S-metolachlor (Dual Magnum, Dual II Magnum) is not used very often because of plant-back restrictions on winter wheat.

Region 2: Very few lentils are grown in Region 2.

Region 3: Canada

- Glyphosate (Roundup Ultra) is applied pre-plant fall or spring for annual weeds, though it is not very effective on buckwheat, prickly lettuce, or any annual weed not yet emerged.

- High rates of glyphosate (Roundup Ultra) are applied for foxtail barley and dandelion control.
- No other pre-plant herbicides are registered.
- Controlling volunteer Round-up Ready canola is a concern.
- Trifluralin (Treflan) and ethalfluralin (Edge, Sonalan) are also used (fall application only), but the requirement for incorporation dries out the seedbed.
- Metribuzin (Sencor, Lexone DF) is used and may be tank-mixed with trifluralin (Treflan) in spring applications only. U.S.
- Glyphosate (Roundup Ultra) is applied pre-plant fall or spring for annual weeds, though it is not very effective on buckwheat, prickly lettuce, or any annual weed not yet emerged.
- High rates of glyphosate (Roundup Ultra) are applied for foxtail barley and dandelion control.
- Imazethapyr (Pursuit) is not used in Montana, due to crop injury potential based on low soil organic matter and low moisture. Pursuit is used in North Dakota.
- North Dakota and Montana do not have registrations for ethalfluralin (Edge, Sonalan).
- S-metolachlor (Dual Magnum, Dual II Magnum) is not registered for North Dakota.
- DNAs are used for a narrow spectrum of broadleaf weeds and some grassy weeds. They do not sufficiently control mustards, kochia, Russian thistle, or wild buckwheat. Downy brome control is important.
- Trifluralin (Treflan) is only labelled for two incorporations, therefore it doesn't fit into low-till or no-till systems.
- Pendimethalin (Prowl) must be incorporated with rain (which doesn't always occur timely).

Region 4: Few lentils are grown in Region 4.

Dry Peas

Region 1:

- Glyphosate (Roundup Ultra) is applied pre-plant fall or spring for annual weeds, though it is not very effective on buckwheat, prickly lettuce, or any annual weed not yet emerged. In effect, nothing is available for buckwheat control.
- Pendimethalin (Prowl) and Ethalfluralin (Edge, Sonalan) are registered for buckwheat control, but herbicide carryover is unacceptable because winter wheat is planted within 6 months.

- Imazethapyr (Pursuit) is most commonly used at the pre-plant stage for annual broadleaf weeds and some grasses but may cause unacceptable crop damage. Reduced rates do not provide effective control. It is weak on wild oat, dog fennel, prickly lettuce and common lambsquarters.
- Pendimethalin (Prowl), applied in combination with imazethapyr (Pursuit), broadens the broadleaf control, but does not control wild oat and other grasses.
- Triallate (Far-Go, Avadex BW), which uses a different mode of action, is used only for wild oat control, but controls no other species and must be incorporated twice.
- Metribuzin (Sencor, Lexone DF) is applied for annual broadleaf weed control. Pre-plant applications are less effective and result in crop damage with heavy moisture (1/2 inch rain within 48 hrs. of application).
- Dimethenamid (Frontier 6EC) was recently registered for annual grasses and broadleaf weeds, but it is weak on dog fennel.
- S-metolachlor (Dual Magnum, Dual II Magnum) is not used very often because of plant-back restrictions on winter wheat.
- Trifluralin (Treflan) is labelled, but is seldom used because of carryover and need for incorporation.
- Pronamide (Kerb) is labelled for pre-plant on winter peas but is not used because of excessive cost and carryover potential.

Region 2:

- Glyphosate (Roundup Ultra) is used, as are ethalfluralin (Edge), trifluralin (Treflan), and metribuzin (Sencor).
- Ethalfluralin (Edge), trifluralin (Treflan), and/or metribuzin (Sencor) can be applied spring or fall, but are best applied in fall. This timing can lead to erosion concerns.
- Amitrole (Amitrol) is registered but has limited use because of the length of the pre-plant interval. It effectively controls dandelion.

Region 3: Canada

- Glyphosate (Roundup Ultra) is used, as are ethalfluralin (Edge), trifluralin (Treflan), and metribuzin (Sencor).
- Ethalfluralin (Edge), trifluralin (Treflan), and/or metribuzin (Sencor) can be applied spring or fall, but are best applied in fall. This timing can lead to erosion concerns.

- Amitrole (Amitrol) is registered but has limited use due to length of pre-plant interval. It effectively controls dandelion. U.S.
- Imazethapyr (Pursuit) is used but it is not labelled for use in North Dakota.
- Sulfentrazone (Spartan), a key herbicide for kochia, Russian thistle, pigweed, and wild buckwheat, is used in Montana, North Dakota and South Dakota under a Section 18 exemption. Preliminary damage reports for dry pea necessitate use rate changes.
- Glyphosate (Roundup Ultra) is used, as are ethalfluralin (Edge), trifluralin (Treflan), and metribuzin (Sencor).
- DNAs are used for a narrow spectrum of broadleaf weeds and some grassy weeds, but do not sufficiently control mustards, kochia, Russian thistle, or wild buckwheat. Downy brome control is important.
- Ethalfluralin (Sonalan) and trifluralin (Treflan) labels require two incorporations, therefore these options don't fit into low-till and no-till systems.
- Pendimethalin (Prowl) and sulfentrazone (Spartan) must be incorporated with rain (which doesn't always occur timely).
- Paraquat is registered but not used due to excessive cost.

Region 4: Canada

- Glyphosate (Roundup Ultra) is used, as are ethalfluralin (Edge), trifluralin (Treflan), and metribuzin (Sencor).
- Ethalfluralin (Edge), trifluralin (Treflan), and/or metribuzin (Sencor) can be applied spring or fall, but are best applied in fall. This timing can lead to erosion concerns.
- Amitrole (Amitrol) is registered but has limited use due to length of pre-plant interval. Good efficacy on dandelion. U.S.
- Imazethapyr (Pursuit) is used but it is not labelled for use in North Dakota.
- Sulfentrazone (Spartan), a key herbicide for kochia, Russian thistle, pigweed, and wild buckwheat, is used under a Section 18 exemption in Montana, North Dakota, and South Dakota. It is labelled for suppression only of wild buckwheat.
- Glyphosate is used, as are ethalfluralin (Edge), trifluralin (Treflan), and metribuzin (Sencor).

- DNAs are used for a narrow spectrum of broadleaf weeds and some grassy weeds, but do not sufficiently control mustards, kochia, Russian thistle, or wild buckwheat. Downy brome control is important.
- Ethalfluralin (Sonalan) and trifluralin (Treflan) are only labelled for two incorporations, therefore these options don't fit into low-till and no-till systems.
- Pendimethalin (Prowl) and sulfentrazone (Spartan) must be incorporated with rain (which doesn't always occur timely).
- Paraquat is registered but not used due to excessive cost.

Critical Needs for Management of Weeds in U.S. and Canadian Pulse Crops Pre-Plant:

Research

U.S. and Canada:

- Develop improved management options at the pre-plant or preemergence stage for buckwheat, kochia, Russian thistle, prickly lettuce, vetch, Round-up Ready canola, and dog fennel
- Investigate herbicides with new modes of action
- Develop methods of application, including timing and rates, to allow existing compounds' use on pulses and rotational crops
- Investigate glyphosate tank-mixes for broader-spectrum weed control
- Investigate carryover of fall-applied phenoxy compounds on pulses
- Improve perennial weed management options
- Investigate interaction of chemical controls with soil organic matter/pH
- Improve winter annual broadleaf control in winter pea and lentil
- Explore pro's and cons of breeding (conventional or otherwise) for herbicide tolerance

Regulatory

Canada:

- Expand DNA labels to include surface application
- More chemical control choices needed; harmonize with U.S. registrations where appropriate data exists in cooperation with registrants
- Explore possibility of growers, registrants, and regulators discussing growers' waiver of liability; acceptance of farmer-generated data (e.g., within the context of pest groupings) U.S.

- Expedite full sulfentrazone (Spartan) registration (all states) for chickpea
- Improve definition of crop groupings
- Distinguish or consider grouping winter and spring biotypes
- Re-evaluate the validity of current field trial regions
- Evaluate the possibility of relief from certain data requirements for minor use situations and label extensions/additions

Education

U.S. and Canada

- *Develop crop profiles*
- *Educate growers on persistence of herbicides in soil*
- *Educate funding sources on the need for more applied research*
- *Conduct field days for regulators to encourage hands-on familiarity* with crops and grower issues
- *Work with registrants on needs.*

Diseases

Chickpeas, lentils, and dry peas are infected by a number of fungi that can cause seed, seedling, and root diseases. These diseases may result in reduced germination, stand establishment, and seed yield and quality. Seed infected shortly after planting may fail to emerge or damp-off soon after emergence. Seedling blight pathogens may also attack adult plants, resulting in reduced leaf and pod development. Fungi causing root rots and seedling blights may affect all pulse crops, predominantly when growing conditions in spring are adverse due to low temperatures, high precipitation, water-logged soils, or drought.

Three primary fungi types are involved in causing seed and root rots: *Rhizoctonia*, *Fusarium*, and *Pythium* species. These fungi can survive in the soil and spread if environmental conditions remain conducive over several seasons. While primarily soilborne, these fungi are also present in crop residue and dust of contaminated seed lots. In some areas soil-borne *Aphanomyces euteiches* causes serious root rot in pea. The pathogen causing ascochyta blight can also over-winter in pulse seed or residue. Ascochyta blight infects all pulses but each crop type is infected by a different *Ascochyta* species. (This is discussed in greater detail in the "Emergence to Harvest" section, below.) Among the pulse crops, chickpea is most severely affected by ascochyta blight and this disease is a key limiting factor to production in North America.

Other significant seed-and residue-borne diseases include sclerotinia (*Sclerotinia sclerotiorum*), botrytis gray mold (*Botrytis cinerea*), and anthracnose of lentil (*Colletotrichum truncatum*).

Cultural Practices

In Canada, it is recommended that pulse growers have their seed tested at an accredited laboratory to determine the percentage of seed-borne disease and germination. Seedborne disease tests are available for ascochyta blight, sclerotinia, and botrytis in all pulses and anthracnose in lentil. Depending on the level of seed-borne disease, the type of pulse crop and percent germination, it may be recommended to not use the seed for planting or to use a fungicide seed treatment.

Pulse growers select vigorous cultivars that are best suited to their climatic region and have the highest level of disease resistance available. There are no cultivars with resistance to diseases caused by Fusarium, Rhizoctonia, Botrytis, Sclerotinia, Pythium or *Aphanomyces* species. Pea cultivars vary in their tolerance to ascochyta blight. There are some lentil cultivars available in Canada with resistance to ascochyta blight and anthracnose. There are some chickpea and lentil cultivars available in the United States with partial resistance to ascochyta blight and breeding efforts are underway to identify improved resistance. Crop rotation with cereals is important to reduce the level of seed- and residue-borne pathogens. However, rotation with non-host crops will not eliminate seed and root rot disease since these pathogens remain viable in the soil for several years and can infect other crops.

Growers plant the same type of pulse crop only once every four years in a field to allow residues to decompose and pathogen numbers to fall off. Other residue management practices such as straw chopping, burial or burning may be considered. In addition, agronomic practices such as seedbed preparation, seed handling, date of planting, row spacing, *Rhizobium* inoculation for nitrogen fixation, and weed control are part of cultural control of disease at the pre-plant stage.

Chemical Controls

In Canada, mefenoxam (Apron XL, Allegiance FL) is registered in pea, chickpea and low tannin lentils as a seed treatment against Pythium species, and in particular, Kabuli chickpea seed is regularly treated. Carbathiin/thiabendazole (Crown) is registered in lentil and chickpea against seed-borne ascochyta blight, but will also protect against other seedand root-rotting fungi. Several fungicides are registered in peas as seed treatments against seed and root rots: Captan/diazinon/lindane (Agrox), captan (Captan FL), thiram (Thiram 75WP), and carbathiin/thiram (Vitaflo-280). Among these, Vitaflow-280 and Thiram 75WP are most commonly used in Canada.

In the United States, Captan (Captan 400) is applied as a seed treatment, usually in combination with mefenoxam (Apron XL). Captan use has decreased over the past few years but continues to be a very effective broad-spectrum fungicide, especially for the control of seedling rot. In the United States, fludioxonil (Maxim) + mefenoxam (Apron XL) is applied as a seed treatment. Fludioxonil provides good control of seedling rot and damping-off from Rhizoctonia and Fusarium, while Pythium is controlled by mefenoxam. No fungicide is available for seed treatment against *Aphanomyces euteiches*. In the United States, seed-borne ascochyta blight is controlled with a Section 18 emergency exemption for thiabendazole (LSP, Mertect). Critical Needs for Management of Diseases in U.S. and Canadian Pulse Crops Pre-Plant:

Research

U.S. and Canada

- Breeding for disease-resistant varieties
- Investigate residue-associated disease management
- Investigate possibility of pre-plant disease control (improved seed treatment, biologicals, etc.)
- Investigate implications of cropping practices (e.g., residue management, rotations) crop sequencing, systems, and rotations on disease incidence and severity
- Investigate effect of soil fertility on disease

Regulatory

U.S.

- Relax regulation of field burning as an alternative during transition U.S. and Canada
- Resolve potential regulatory issues with some fungicide seed treatments

Education

U.S. and Canada

- Educate research funding sources about the importance of supporting plant breeding

- Educate growers about residue/stubble management alternatives.
- Educate growers on disease management

Pre-Emergence

After planting, pulse crop beds are rolled to smooth the soil surface. This field operation improves the harvesting of low-hanging pulse pods by reducing harvest losses and breakage of sickle section and guards, and improves harvest rate. Rolling is done anytime from immediately after seeding up to the 5- to 7-node stage in lentils and up to the 5-leaf stage in peas. Chickpeas are rolled prior to emergence only.

Weeds

Pulse crops are weak competitors with weeds. Limited herbicide choices and increased annual and perennial weed densities, resulting from volunteer crops as weeds and the increased importance of winter annuals and secondary weed competitors, present special weed control challenges. Weed control at the pre-emergence stage is important to help encourage vigorous seedling growth for effective competition.

Chemical Controls

Metribuzin (Sencor) is applied pre-emergent on a few acres of lentils when weed control is not achieved by the pre-plant incorporated herbicides. It can cause crop stress under high-precipitation weather conditions. Other herbicides that are applied pre-emergent include dimethenamid (Frontier 6EC), imazethapyr (Pursuit), and s-metolachlor (Dual Magnum, Dual II Magnum).

Critical Needs for Management of Weeds in U.S. and Canadian Pulse Crops at Pre-Emergence:

Research

U.S. and Canada

- Determine optimal timing and method of application of weed control product
- Investigate pre-emergent annual broadleaf control in chickpeas and lentils
- Investigate mechanical methods of weed control
- Research expansion of herbicide use during the fall for winter legumes.

Regulatory

U. S. and Canada

- Expand Pursuit label, harmonization for all 3 crops
- Register label for sulfentrazone (Spartan) on dry peas and chickpeas
- Work with registrants to maintain active ingredient registrations
- Note that registrant response time is slow due to consolidation of registrants.

Canada

- Register label for ethalfluralin (Edge) on chickpea
- Register label for trifluralin (Treflan) on chickpea.

Education

U.S. and Canada

- Educate registrants on our registration priorities; work with them to maintain active ingredient registrations.

Diseases

Seedling blights, root rots, botrytis gray mold, and seed-borne ascochyta blight can infect pulse crops at pre-emergence. Critical Needs for Management of Diseases in U.S. and Canadian Pulse Crops at Pre-Emergence:

Research

U.S. and Canada

- Screen germplasm for resistance to damping off and root rot (Pythium and Rhizoctonia).

Regulatory

- *None.*

Education

- *None.*

Insects

Insects are not controlled at the pre-emergent stage.

Chapter

12

Future Requirements for Insecticides and Application Equipment

Insecticides

Candidate compounds for tsetse control from commercial companies, research institutes and other sources should continue to be assessed for insecticidal activity in laboratory tests as a basis for selection for field evaluation. This selection should take into consideration not only toxicity to tsetse flies, but also existence of a suitable formulation, availability and cost, toxicology and possible effects on non-target organisms. Some of these factors are listed for new compounds already regarded as possible replacement chemicals on the basis of toxicity to flies. The availabilities mentioned are of the order of 1g to 100g for laboratory testing, 1 kg to 100 kg for field trials, and in normal commercial quantities for those which are readily available and apply to the situation as far as it is known at the present time.

Highest priority should be given to field evaluation of the synthetic pyrethroids, NRDC 143 and NRDC 161, which are outstanding in their toxicity to tsetse flies and their persistence in spray residues and could be used in both aerial and ground sprays. NRDC 161 is the most potent insecticide known against many insects but also has a high mammalian toxicity with a site of action which differs from other pyrethroids. It is important therefore that further information should be obtained as soon as possible on hazards to other animals, although it should also be noted that the very powerful insecticidal action means that it can be used at lower concentrations and dosages than the other candidates. The organophosphorus compounds and carbamates listed have good to moderate toxicities to the flies as measured by topical application and can be divided into those which are likely to have a short term persistence only (for example, fenthion, naled, bromophos, propoxur) and those which could have a longer residual action (for example, azamethiphos, tetrachlorvinphos, jodfenphos, bendiocarb). Bromophos from the former group and tetrachlorvinphos from the latter have already been examined in field trials and found to be not very promising.

The susceptibility of different species and strains of tsetse to endosulfan and dieldrin and to representatives of other chemical classes that are likely to be used in field operations should be measured in different parts of Africa. This will provide guidance for variations in dosages required and also baseline data for a systematic monitoring programme for resistance. A simple standard method should be adopted; the most suitable is probably the WHO method for topical application, but field performance should also be studied by methods utilizing tarsal contact with weathered deposits on sprayed surfaces (e.g. bark) if persistence is to be investigated.

Residual Applications to Vegetation at High Volume Rates

Solutions are not usually employed because the solvent is expensive compared with water, but they were used in early trials in East Africa with DDT or more recently in Nigeria with dieldrin. The cheaper solvents, such as kerosene, which can be used at high volumes, evaporate rapidly from deposits and do not influence persistence or activity of the insecticide by their presence. They can modify the initial distribution of insecticide by moving it into bark or leaves before they evaporate.

Emulsions or suspensions are much commoner systems for insecticide dispersal, and the two insecticides, DDT and dieldrin, are invariably used in these forms in the larger control schemes, for example in Nigeria, Uganda, Tanzania, Kenya and Zambia. Suspensions are prepared from water dispersible powders which contain insecticide and inert diluent particles and surfactants and give dry deposits since the water evaporates very rapidly. Biological potency depends upon the particle size distribution of the active ingredient, its physical state and the surface which has been treated. Finer particles are picked up more readily by resting insects, while the adhesive properties of surfactant residues reduce pick up on the less porous surfaces. Those powders which meet WHO specifications have about the optimum particle size distribution. Since the insecticide is in solution in emulsion concentrates, or in the emulsions prepared from these, the physical forms of the deposit after solvent and water have evaporated will be controlled by the crystallization properties of the insecticide and, again, the sprayed surface since this crystallization occurs more readily on some surfaces than others.

The persistence of the deposits resulting from the use of powders or emulsions will be controlled by the physical properties of the active material (e.g. dieldrin is more volatile than DDT), by climatic variables and by the surface sprayed. Thus, water-dispersible powder deposits have been found more effective in hot drier areas than those from emulsions, but the reverse is true in cooler areas with higher rainfall. The powder residues are more easily washed away by rain than those from emulsions. Another important factor is the intrinsic toxicity of a compound to the flies. For example, dieldrin is more potent than DDT and deposits can be reduced by a larger amount but still be more effective. All the candidates for replacement insecticides in tsetse control are available, or could be made available, as water-dispersible powders and emulsion concentrates. A few, such as pirimetaphos, azamethiphos, tetrachlorvinphos and bendiocarb, have a low solubility in the common solvents used to prepare emulsion concentrates, but this could be improved with known co-solvents. Conversely, their low solubilities mean that they usually make good water-dispersible powders.

Aerial Application at Ultra-low Volume Rates

This usually requires that the formulation should be a solution of high concentration, say 10% upwards, since it is important that a lethal dose is present in one or very few droplets. Although this criterion applies to most insecticides it is possible that with exceptionally toxic compounds, such as NRDC 161, the concentration can be reduced to something of the order of 1%. For these solutions the vital component is the solvent or, more often, solvents. Their properties control the initial atomization, the reduction in droplet size during passage to the target and the efficiency with which the insecticide passes into the insect, either from droplets on the cuticle or from droplets on surfaces on which the insect rests. They can also influence the persistence of insecticides in spray deposits when they are of low volatility themselves.

A discussion of all these interactions between insecticide, solvent, insect and surface would be too long for this report. Briefly, however, the best solvent system probably consists of a non-volatile component such as a vegetable oil, plasticizing ester or Dutrex-type petroleum extract combined with a volatile solvent which may be lost during application but ensures that the physical properties of the blend are appropriate for the atomization process. The non-volatile solvent is a less efficient carrier for insecticide to the design and execution of substantial antitsetse activities.

Specific Consideration

The natural limits of the tsetse population to be attacked should be determined before control operations are initiated. Surveys should be undertaken with care to determine population limits and maximum dispersal capabilities at the most favourable seasons of the year. Low density populations require the development of new sampling techniques. Restricted application of residual insecticide requires prior knowledge of those plant communities with which a large proportion of the tsetse population is associated, especially at unfavourable seasons of the year ("discriminative" phase). Resting sites utilized by the flies throughout the day and night in these plant communities should be determined ("selective" phase). Resting sites can be confined to a small proportion of the overall habitat, and very restricted use of insecticide is possible under such circumstances, especially when applied by ground spray teams applying the principles of "discrimination" and "selection". Findings from one area are not necessarily applicable elsewhere within the range of a species and such studies are necessary wherever control operations are to be undertaken.

Less detailed knowledge of the ecology of the species to be attacked is necessary when blanket applications of non-residual insecticide are to be employed. In order to determine the optimum number and interval between application it is, however, necessary to know the duration of the pupal period throughout the time when spraying is to be conducted; particular attention should be paid to the coolest breeding sites where the period will be most prolonged. It is the maximum period and not the mean which is most significant. For the same reason, it is also necessary to determine the length of time between the emergence of a female fly and the production of its first lava throughout the spray period, particularly at the shortest period.

At the conclusion of control operations, detailed surveys are again required to determine whether or not eradication has been achieved.

The detection of tsetse flies at very low population densities is difficult and inefficient and the development of appropriate methods is perhaps the most pressing need in studies of Glossina ecology. Now efficient devices for sampling tsetse populations have been developed; the possibility of adapting these to detect low density populations requires urgent investigation.

Environmental Effects

If land is to be opened up for the purpose of providing food resources, then it must be recognized that the agricultural activities themselves which will ensue will bring about a far greater environmental change than initial applications of insecticides for tsetse control. Furthermore, the agricultural changes will require greater amounts of more persistent insecticides for crop protection than will be used in public health. Accordingly, planners of such large programmes should approach the problem of environmental modification from a realistic viewpoint. Governments which seek the long-term objective of increased agricultural yield cannot at the same time criticize too severely the short-term use of insecticides for vector control. The alternatives to the use of insecticides for tsetse control are the direct killing of game and habitat clearance. The environmental effects of these operations are probably greater than the use of the insecticides. Although genetic control of Glossina is a distinct feasibility, a great deal of work will have to be done to determine its operational value. Despite recent publications on the possibilities of biological control of Glossina, it is considered to have a remote feasibility.

For tsetse control certain areas of wooded savannah, riverine forests and forests proper, including the ecotones between such systems, will be treated. The first step would be to map or classify the major ecosystems that will be covered by treatment and, based upon existing literature, to provide a preliminary listing and structure of the major animal groups in each. From this, a series of indicator organisms can be derived and their availability and relative abundance in each of those main ecosystems approximated and checked by spot surveys. Once spraying operations have been decided upon and the toxicity of the insecticide (s) selected determined for major groups, i.e. mammalian, avian, fish, etc., a sampling procedure for the indicator species population levels can be established. There are numerous adequate procedures for determining vertebrate population levels depending on availability of manpower and the precision desired. Other major groups to be sampled would be non-target forest insects, as well as soil organisms, especially Annalida and arthropods.

Physical measurements to determine the distribution and quantity of insecticides applied by aerial spraying in sample habitats should also be made. Initially, this should be done in a wooded area at a canopy level, at intermediate tree level, on ground vegetation and, ultimately, in the soil. Again, details can be determined for each of the habitat types.

Future Research Requirements

Recent progress in the use of fixed-and rotary-wing aircraft and ground equipment for insecticide applications needs to be exploited in a wider variety of situations and new insecticide formulations require field evaluation. Thus, field studies to evaluate new techniques and chemicals should be extended to situations where treatments have been difficult or impossible in the past, as in certain riverine, moist savanna and forest zones. In some situations where control has been successful alternative methods are desirable as, for example, when economic or environmental aspects must be taken into account.

Laboratory Evaluation of New Insecticides and Formulations

CORP is engaged in a continuing programme for the laboratory testing and evaluation of candidate insecticides for toxicity to tsetse flies in collaboration with WHO. Comparative toxicities of compounds supplied by industry, research institutes, universities and other sources are determined by a standard method of topical application of solutions to teneral G. austeni obtained from pupae*. The results are published from time to time and indicate the most promising insecticides for space sprays. Other properties, such as volatility, solubility and stability are also measured to indicate suitability of compounds for residual activity. A research programme on formulations of insecticides for residual activity on vegetation is also currently in progress. A spray tower has been constructed for the simulation of ULV conditions; that is, the application of concentrated solutions of 10% or more active ingredient at low volume rates and controlled droplet size. Observations are made on the physical state of deposits on leaves and of the persistence of the active ingredient on leaves. The persistence and availability of the active ingredient is measured by bioassay and chemically by GLC analysis.

Field Studies of New Insecticides and Techniques

Ground Applications of Residual Insecticides

Persistent chlorinated hydrocarbon insecticides, e.g. DDT, dieldrin and BHC are in current use for tsetse fly control. Insecticides are applied selectively to the known resting sites of flies, e.g. the trunks of trees and the undersides of certain branches or to leaves in the canopy and are effective when applied to riverine and savanna vegetation using ground spraying machines. Insecticides which are environmentally more acceptable and less toxic to cattle and humans would be evaluated for use by this method. Trials with these insecticides should include dieldrin as a reference treatment. Details of proposed trials are shown in Appendix II.

Ground Applications of Non-Residual Insecticides

Hand carried equipment for applying aerosols for tsetse fly control has been used in various parts of Africa, but the method has never been developed to any extent. Equipment applying DDT as fogs has been tried but there is a need to evaluate alternative less persistent insecticides applied as ULV or as dilute formulations suitable for fogging. These methods are unlikely to lead to permanent eradication, but there are requirements to reduce fly numbers rapidly in emergency sleeping sickness situations, in areas of human habitation or of tourist interest. It would be necessary to study sequential applications of insecticide if long-term control was required. Applications of insecticide from boats would have to be considered for access to difficult riverine areas. Details are given in Appendix III.

Rotary Wing Aircraft Applications of Residual Insecticides

Persistent chlorinated hydrocarbon insecticides, e.g. DDT, dieldrin and endosulfan have recently been shown to control tsetse flies effectively in West Africa by discriminative treatments of aerial sprays applied by helicopter to forest islands, drainage lines, ecotones and stretches of river where the vegetation is suitable. Since aerial applications are necessarily more diffuse than hand spraying, it is vital to evaluate alternative insecticides which are environmentally safer and less toxic to man and cattle. It is also important to examine closely the economics of this method and the possibility of using different formulations for greater efficiency. Details are given in Appendix IV.

Rotary Wing Aircraft Applications of Non-residual Insecticides

The control of tsetse flies along large stretches of rivers where ground access is difficult has always presented problems. Fixed-wing aircraft do not have the manoeuverability to follow the course of the riverine vegetation and residual spraying from the air causes greater environmental side effects. There is, therefore, a need to devise a non-residual spraying technique using environmentally safer insecticides for the rapid reduction of fly, especially in areas where sleeping sickness is epidemic.

Medium Fixed Wing Aircraft Applications of Non-residual Insecticides

The technique of applying aerosols from fixed-wing aircraft for tsetse fly control has been well tried and has been shown to be effective with a variety of insecticides in field experiments and in large airspray operations but requires additional development work. Endosulfan is currently used at rates of 6–14 g/ha per treatment. There may be situations where this insecticide is a hazard, e.g. around rivers or lakes, and it would be worthwhile assessing alternative insecticides. Details are given in Appendix VI.

Insecticide Application Equipment

The compressed air sprayer, the quart-sized sprayer, the paintbrush, and the bulb duster are the best equipment to use for applying household insecticides. Each type has its pluses and minuses. Careful study of your insect-control problem will help you choose the best equipment for the job.

Compressed Air Sprayer

The water capacity of a compressed air sprayer is usually one to four gallons. Air is pumped into the tank, forcing the spray out when the nozzle is opened. It is ideal for outdoor application of wettable powders and emulsions. Its use indoors is limited if a lot of water is applied with the insecticide. Shake the sprayer frequently when using.

Quart-sized Sprayer

The quart-sized sprayer is also a compressed-air type, but air must be pumped into it continuously while in use. It can be used satisfactorily with emulsions and solutions but not wettable powders. Use it both indoors and outdoors for treating small areas. However, where high pressure is needed, these have limited use.

Paintbrush

Use an inexpensive paintbrush to apply insecticide solutions to baseboards, screens, and similar areas inside buildings. A light film is usually sufficient.

Bulb Duster

This apparatus is used for applying dusts to cracks, crevices, and other voids. The dust is placed into the bulb or bellows; squeezing the device forces the dust out the nozzle and into the void to be treated. A little bit of practice (with plaster of Paris or flour) will help you perfect your dusting technique. Best results are achieved when dusts are applied in a thin, even layer. If you can easily see the dust on a surface, you have used too much.

Application Equipment

Many European corn borer control treatments are applied aerially. In areas with a history of European corn borer problems, granules, if available for an approved insecticide, are the preferred formulation, particularly for control during the whorl stage of corn development. Fortunately, most aerial operators are equipped to apply either a granule or a liquid formulation. Although not true of all compounds, efficacy comparisons between granules and liquids of the same compound applied to whorl-stage corn usually favor granules. Granules have additional advantages: (1) no mixing or water is needed, (2) there is less drift, (3) personal hazard is usually less, and (4) there is less hazard to honeybees and other beneficial insects. Granule insecticides also better control larvae during the tasseling and silking stages of corn development. Granules roll down plant leaves easily and tend to concentrate close to larvae feeding in the whorl. However, liquid insecticides provide good control and are appropriate when conditions are unfavorable for granules.

High-clearance equipment, when available, can be modified to apply either granules or liquids. To control leaf feeding, granules must be directed into the plant whorl. To control sheath and collar feeding, granules must be directed over the entire plant and into leaf axils. This can be done by using 7- to 10-inch (18- to 25-cm) spreader banders attached to delivery tubes extending from each insecticide box. Be careful to follow label directions closely, especially to minimize effects on birds and other wildlife.

For liquid applications to reduce leaf feeding, use one or two nozzles per row. Adjust the nozzles to deliver the spray above and into the whorl. To control sheath, collar, and ear feeding, use four nozzles per row and direct the spray at the ear zone. Thorough coverage is important. Ground applications should be applied with a minimum of 10 gallons of water per acre. To control whorl feeding, granule or liquid delivery systems can be mounted on a tool bar and used with cultivation equipment. However, corn usually is too tall and already cultivated when the decision to control whorl feeding is made. Although there are advantages to ground application, equipment availability and time requirements for application to large acreages often force using aerial equipment. In areas where insecticides commonly are applied at planting or during cultivation to control rootworm larvae, systemic insecticide application has been inconsistent in controlling whorl-feeding European corn borer and is not recommended.

Insecticide application through irrigation water (insectigation), particularly with overhead centre pivot irrigation systems, is an effective method and is becoming common in some irrigated areas of the Corn Belt. Because the method is relatively new, technological and environmental tolerances are changing. Before using this application method, consult your local Cooperative Extension Service and regulatory agencies. Some states have very specific rules and regulations regarding the injection of chemicals into irrigation water. One of the primary concerns of insectigation is potential underground water contamination. If insectigation is used, comply with all laws and regulations, including those promulgated by the Environmental Protection Agency, the manufacturer of the insecticide, and local regulatory agencies. At a minimum, make sure the well is equipped with a check valve to eliminate the possibility of reverse flow, or back siphoning, in case the system loses pressure and shuts down during the process. The check valve will minimize the possibility of contamination of the well and underground water.

Commercial Applications of Insecticides and Miticides in the Green Industry

Adjuvants are substances added to a pesticide formulation by the manufacturer or tank mix by the applicator to improve its performance, ease of application, or safety. Some of the most common adjuvants are surfactants, which change the dispersing, spreading, and wetting properties of spray droplets. Spreaders are adjuvants that allow the pesticide to form an even coating over treated surfaces, while stickers help the pesticide to stay on or adhere to the foliage. Other adjuvants may reduce the foaming of mixtures that need agitation or reduce drift during application. Buffers allow pesticides to be mixed with water or other pesticides of different acidity or alkalinity. For example, a buffer should be added to a spray tank if the water pH is greater than 7, to maintain the pH near 6.5. Compatibility agents aid in combining and mixing different pesticides effectively.

Insecticides, although used to maintain healthy plants and reduce pest populations, have the potential to damage sensitive plants. Phytotoxicity often appears as a marginal burn, chlorosis, spotting, distortion, or abnormal growth of leaves. Although any part of the plant may be affected, the new growth is most likely to show damage. For example, an improper tank mixture of incompatible products may cause phytotoxicity. Consult a compatibility chart and manufacturer labels before mixing materials. In addition, pesticides are less likely to damage plants if treatments are done in the early morning. Foliage will be dry before temperatures get too hot and sunlight magnifies through spray droplets. In general, powders can be mixed together, and a powder or soluble can be added to an emulsifiable concentrate. Mixes should be of compounds either of the same class or the same brand, if possible. Regardless of the pesticide or mixture of pesticides used, it is strongly recommended that the effects be evaluated on a few plants, under your particular conditions, before treating all plants.

A systemic insecticide is a chemical compound that is absorbed by the host plant, translocated throughout its tissues and makes the host toxic to certain insect and mite pests. Several systemic insecticides are taken up from the soil by the roots of plants and translocated throughout the plant tissues; others can be absorbed by foliage or stem sprays. Systemic insecticides have been effective primarily in controlling small sucking pests, including aphids, whiteflies, scales, mealybugs, lace bugs and spider mites. Historically, they have not given satisfactory control of chewing insects.

Dry, granular applications may be used to control soil-dwelling insects. These insecticides are often applied with a walk-behind rotary spreader. To ensure even distribution over the soil surface, applicators often walk back and forth in one direction (e.g., north-south) until the area is covered, then criss-cross in the other direction (e.g., east-west). Granular products often need to be watered into the soil to prevent binding with plant material and to move the insecticide into the thatch or soil. Granules may cause phytotoxicity if applied to wet foliage, especially on ornamentals. Carefully calibrate spreaders so they do not grind the granules or create dust because of the increased potential for applicator exposure or inhalation of the product.

Spray applications are often done using flowable, emulsifiable concentrates, and other insecticide formulations. Spray equipment varies from hand-held pumps and backpack sprayers to high-pressure hydraulic sprayers. Thorough coverage is important to gain adequate control of insect pests. Some labels suggest applying sprays until foliage is misted or until runoff is achieved. Sufficient pressure is necessary to turn leaves and reach insects feeding on the underside of leaves. For best results, many insecticides need to be mechanically agitated in the spray tank to remain in solution. Drench applications may be useful against root-feeders, wood- or phloem-borers, or sap-feeders. The insecticide (e.g., various neonicotinoids) is sprayed or poured within the plant's drip-line, and the plant's root system takes it up. The insecticide is then translocated systemically throughout the plant and kills the pests as they feed. Soil drenches are best applied a short time before pests are expected to be active. Chemigation is the application of systemic products through an irrigation system. For best results, apply the products toward the end of the irrigation period. Do not apply insecticides using chemigation to plants under water stress, or follow application with excessive amounts of water. Some insecticides are not labelled for use through chemigation.

Trunk injections directly apply insecticides (e.g., imidacloprid, abamectin, emamectin benzoate, bidrin) into a tree for faster translocation. Several systems for macro- or microinfusion exist, including Mauget, Arborjet, and Wedgle. Small, shallow holes are drilled near the base of the tree trunk. Tubes are inserted into the holes and the insecticide is injected under pressure into the trees. Trunk injections reduce drift and exposure, but may be more expensive and labour-intensive, and wounds are created in the tree that may take time to heal and allow entry by pathogens.

Subsurface application of insecticides is used for control of soil-dwelling pests such as mole crickets in turfgrass. This includes high-pressure liquid injection or slit-injection of granular or liquid products. The goal is to place the insecticide in the soil where the pest lives. This technique has several advantages. Because the insecticide is delivered to the target zone, less is bound in the thatch layer. Losses from volatilization or degradation in sunlight are reduced. Less residue occurs on the surface, which reduces the hazard to workers, turfgrass users, or wildlife. Spray drift and odor are also reduced. Potentially lower insecticide rates may be needed, compared to surface treatments. Turfgrass looks torn up after treatments are applied, but thatch is reduced and the grass recovers. Disadvantages include needing to buy expensive application equipment or contract out for applications.

Smokes, aerosols, and thermal fogs may be used in greenhouses if temperatures are cool. However, these formulations are not recommended for use during hot periods. Little research has been conducted with these products on tropical foliage plants.

Important

Federal and Florida Laws state that all pesticides must be handled and applied in strict accordance with the label and worker protection standards (re-entry times, protective clothing, etc.). According to the Federal Insecticide, Fungicide and Rodenticide Act (FIFRA), the plant must be listed on the label. FIFRA does allow growers to apply pesticides against pests not specifically listed on the label. However, if a pesticide is used in a greenhouse or interiorscape, it must be labelled for such use. To the best of our knowledge, the suggested pesticides listed in this management guide are labelled on the crops and sites discussed. However, pesticide labels differ widely in plant and pest listings. Some are broad ornamental labels and others specifically list certain plants for which the pesticide is labelled.

Read the entire label carefully, including the small print, before opening the container. Avoid drift of pesticides to nearby areas or to plants that may be eaten by people or animals. Do not allow pesticides to contaminate wells, ponds, or streams. Store pesticides in their original labelled containers, in a locked room with metal shelves. Rinse empty containers several times with water and pour rinsing in spray tank. Puncture and dispose of empty containers promptly and safely. Keep the telephone number and address of the nearest Poison Control Centre listed in a convenient location in case of an accidental pesticide poisoning. Also, keep the labels and Material Safety and Data Sheets (MSDS) of all pesticides that are on the premises. If a poisoning occurs, show the pesticide label to the physician. If a product is labelled for use specifically in Florida (special local need), obtain a copy of the supplemental label from the supplier when the product is purchased.

Because cultural conditions vary widely in the Green Industry, no single pest control program can be suggested. Chemical control of insects is likely only when the correct material is applied in the correct manner to a susceptible stage of the pest. Maintenance or so-called preventive sprays may be applied every 1 to 3 weeks, depending upon the pest, time of year, and residual length of a pesticide. We strongly suggest that a scouting program be conducted and insecticides be applied only when truly needed.

Pesticide Application

Pesticide application refers to the practical way in which pesticides, (including herbicides, fungicides, insecticides, or nematode control agents) are delivered to their *biological targets* (*e.g.* pest organism, crop or other plant). Public concern about the use of pesticides has high-lighted the need to make this process as efficient as possible, in order to minimise their release into the environment and human exposure (including operators, bystanders and consumers of produce). The practice of pest management by the rational application of pesticides is supremely multi-disciplinary, combining many aspects of biology and chemistry with: agronomy, engineering, meteorology, socio-economics and public health, together with newer disciplines such as biotechnology and information science.

Seed Treatments

Seed treatments can achieve exceptionally high efficiencies, in terms of effective dose-transfer to a crop. Pesticides are applied to the seed prior to planting, in the form of a seed treatment, or coating, to protect against soil-borne risks to the plant; additionally, these coatings can provide supplemental chemicals and nutrients designed to encourage growth. A typical seed coating can include a nutrient layer—containing nitrogen, phosphorus, and potassium, a rhizobial layer—containing symbiotic bacteria and other beneficial micro-organisms, and a fungicide (or other chemical) layer to make the seed less vulnerable to pests..

Spray Application

One of the more common forms of pesticide application, especially in conventional agriculture, is the use of mechanical sprayers. Hydraulic sprayers consists of a tank, a pump, a lance (for single nozzles) or boom, and a nozzle (or multiple nozzles). Sprayers convert a pesticide formulation, often containing a mixture of water (or another liquid chemical carrier, such as fertilizer) and chemical, into droplets, which can be large rain-type drops or tiny almost-invisible particles. This conversion is accomplished by forcing the spray mixture through a spray nozzle under pressure. The size of droplets can be altered through the use of different nozzle sizes, or by altering the pressure under which it is forced, or a combination of both. Large droplets have the advantage of being less susceptible to spray drift, but require more water per unit of land covered. Small droplets are able to static electricity to maximize contact with a target organism, but require very still wind conditions.

Spraying Pre- and Post-emergent Crops

Traditional agricultural crop pesticides can either be applied pre-emergent or post-emergent, a term referring to the germination status of the plant. Pre-emergent pesticide application, in conventional agriculture, attempts to reduce competitive pressure on newly germinated plants by removing undesirable organisms and maximizing the amount of water, soil nutrients, and sunlight available for the crop. An example of pre-emergent pesticide application is atrazine application for corn. Similarly, glyphosate mixtures are often applied pre-emergent on agricultural fields to remove early-germinating weeds and prepare for subsequent crops. Pre-emergent application equipment often has large, wide tires designed to float on soft soil, minimizing both soil compaction and damage to planted (but not yet emerged) crops. A three-wheel application machine, such as the one pictured on the right, is designed so that tires do not follow the same path, minimizing the creation of ruts in the field and limiting sub-soil damage.

Post-emergent pesticide application requires the use of specific chemicals chosen minimize harm to the desirable target organism. An example is 2,4-Dichlorophenoxyacetic acid, which will injure broadleaf weeds (dicots) but leave behind grasses (monocots). Such a chemical has been used extensively on wheat crops, for example. A number of companies have also created genetically-modified organisms that are resistant to various pesticides. Examples include glyphosate-resistant soybeans and Bt maize, which change the types of formulations involved in addressing post-emergent pesticide pressure. It is important to also note that even given appropriate chemical choices, high ambient temperatures or other environmental influences, can allow the non-targeted desirable organism to be damaged during application. As plants have already germinated, post-emergent pesticide application necessitates limited field contact in order to minimize losses due to crop and soil damage. Typical industrial application equipment will utilize very tall and narrow tires and combine this with a sprayer body which can be raised and lowered depending on crop height. These sprayers usually carry the label 'high-clearance' as they can rise over growing crops, although usually not much more than 1 or 2 meters high. In addition, these sprayers often have very wide booms in order to minimize the number of passes required over a field, again designed to limit crop damage and maximize efficiency. In industrial agriculture, spray booms 120 feet (40 meters) wide are not uncommon, especially in prairie agriculture with large, flat fields. Related to this, aerial pesticide application is a method of top dressing a pesticide to an emerged crop which eliminates physical contact with soil and crops.

Air Blast sprayers, also known as air-assisted or mist sprayers, are often used for tall crops, such as tree fruit, where boom sprayers and aerial application would be ineffective. These types of sprayers can only be used where overspray—spray drift—is less of a concern, either through the choice of chemical which does not have undesirable effects on other desirable organisms, or by adequate buffer distance. These can be used for insects, weeds, and other pests to crops, humans, and animals. Air blast sprayers inject liquid into a fast-moving stream of air, breaking down large droplets into smaller particles by introducing a small amount of liquid into a fast-moving stream of air. Foggers fulfill a similar role to mist sprayers in producing particles of very small size, but use a different method. Whereas mist sprayers create a high-speed stream of air which can travel significant distances, foggers use a piston or bellows to create a stagnant area of pesticide that is often used for enclosed areas, such as houses and animal shelters.

Spraying Inefficiencies

In order to better understand the cause of the spray inefficiency, it is useful to reflect on the implications of the large range of droplet sizes produced by typical (hydraulic) spray nozzles. This has long been recognized to be one of the most important concepts in spray application (*e.g.* Himel, 1969), bringing about enormous variations in the properties of droplets.

Historically, dose-transfer to the biological target (*i.e.* the pest) has been shown to be inefficient. However, relating "ideal" deposits with biological effect is fraught with difficulty), but in spite of Hislop's misgivings about detail, there have been several demonstrations that massive amounts of pesticides are wasted by run-off from the crop and into the soil, in a process called endo-drift. This is a less familiar form of pesticide drift, with exo-drift causing much greater public concern. Pesticides are conventionally applied using hydraulic atomisers, either on hand-held sprayers or tractor booms, where formulations are mixed into high volumes of water.

Different droplet sizes have dramatically different dispersal characteristics, and are subject to complex macro- and micro-climatic interactions (Bache & Johnstone, 1992). Greatly simplifying these interactions in terms of droplet size and wind speed, Craymer & Boyle concluded that there are essentially three sets of conditions under which droplets move from the nozzle to the target. These are where:

- *Sedimentation dominates:* typically larger (>100 μm) droplets applied at low wind-speeds; droplets above this size are appropriate for minimising drift contamination by herbicides.

- *Turbulent eddies dominate:* typically small droplets (<50 μm) that are usually considered most appropriate for targeting flying insects, unless an electrostatic charge is also present that provides the necessary force to attract droplets to foliage. (NB: the latter effects only operate at very short distances, typically under 10 mm.)
- Intermediate conditions where both sedimentation and drift effects are important. Most agricultural insecticide and fungicide spraying is optimised by using relatively small (say 50-150 μm) droplets in order to maximize "coverage" (droplets per unit area), but are also subject to drift.

Improved Targeting

In the 1970s and 1980s improved application technologies such as controlled droplet application (CDA) received extensive research interest, but commercial uptake has been disappointing. By controlling droplet size, ultra-low volume (ULV) or very low volume (VLV) application rates of pesticidal mixtures can achieve similar (or sometimes better) biological results by improved timing and dose-transfer to the biological target (*i.e.* pest).

No atomizer has been developed able to produce uniform (monodisperse) droplets, but rotary (spinning disc and cage) atomizers usually produce a more uniform droplet size spectrum than conventional hydraulic nozzles. Other efficient application techniques include: banding, baiting, specific granule placement, seed treatments and weed wiping.

CDA is good examples of a rational pesticide use (RPU) technology (Bateman, 2003), but unfortunately has been unfashionable with public funding bodies since the early 1990s, with many believing that all pesticide development should be the responsibility of pesticide manufacturers.

On the other hand, pesticide companies are unlikely widely to promote better targeting and thus reduced pesticide sales, unless they can benefit by adding value to products in some other way. RPU contrasts dramatically with the promotion of pesticides, and many agrochemical concerns, have equally become aware that product stewardship provides better long-term profitability than high pressure salesmanship of a dwindling number of new "silver bullet" molecules. RPU may therefore provide an appropriate framework for collaboration between many of the stake-holders in crop protection.

Other Application Methods

- Granule application equipment
- Dust application
- Build a wheeled caddy for pump garden sprayer.

Application Methods for Household Insecticides

Pest management in the home begins with restricting the availability to insects of three vital commodities: shelter, water and food. If insects become a problem despite such measures, it becomes to control them using the safest possible methods, targeting the active ingredient to the particular pest.

Insect repellent, referred to as "bug spray", comes in a plastic bottle or aerosol can. Applied to clothing, arms, legs, and other extremities, the use of these products will tend to ward off nearby insects. This is not an insecticide.

Insecticide used for killing pests—most often insects, and arachnids—primarily comes in an aerosol can, and is sprayed directly on the insect or its nest as a means of killing it. Fly sprays will kill house flies, blowflies, ants, cockroaches and other insects and also spiders. Other preparations are granules or liquids that are formulated with bait that is eaten by insects. For many household pests bait traps are available that contain the pesticide and either pheromone or food baits. Crack and crevice sprays are applied into and around openings in houses such as baseboards and plumbing. Pesticides to control termites are often injected into and around the foundations of homes.

Active ingredients of many household insecticides include permethrin and tetramethrin, which act on the nervous system of insects and arachnids.

Bug sprays should be used in well ventilated areas only, as the chemicals contained in the aerosol and most insecticides can be harmful or deadly to humans and pets. All insecticide products including solids, baits and bait traps should be applied such that they are out of reach of wildlife, pets and children.

Bibliography

Aaboe, Asger: *Episodes from the Early History of Mathematics*, New York, 1964.

Andrews, J.: *The Domesticated Capsicums*, University of Texas Press. 1995.

Andrews, L.: *Citrus Production - Orange*, St. Augustine, Trinidad and Tobago, 1990

Bag, A. K.: *Mathematics in Ancient and Medieval India*, Varanasi: Chaukhambha Orientalia, 1979.

Banki, L.: *Bioassay of Pesticides in the Laboratory,* Akademiai Kiado, Budapest, 1978.

Baruah, Akhil: *Advanced Morphology of Angiosperms*, Aavishkar, Delhi, 2008.

Chambers, R.: *Rural Development: Putting the Last First*, London, Longman, 1983.

Clawson, Calvin: *The Mathematical Traveller*, New York, Plenum Press, 1994.

Crump, Thomas: *The Anthropology of Numbers*, Cambridge, Cambridge University Press, 1990.

Dabholkar, A.R.: *General Plant Breeding*, Concept, Delhi, 2006.

Dudley, E.: *The Critical Villager: Beyond Community Participation*, London, Routledge, 1993.

Featherly, H. I.: *Taxonomic Terminology of the Higher Plants*, USA, Iowa State College Press, 1954.

Fortuner, R.: *Nematode Identification and Expert System Technology,* New York, Plenum Press, 1988.

Godden, G.: *Growing Citrus Trees*, Australia, Lothian Publishing Company Pvt. Ltd., 1988.

Gowen, S.: *Banana and Plantains*, London, Chapman Hall, 1996.

Herklots, G. A. C.: *Vegetables in South East Asia*, London, George Allen & Unwin Ltd., 1972.

Huaman, Z.: *Descriptors for Sweet Potato*, Rome, International Board for Plant Genetic Resources, 1991.

Jeffers, P.: *Evaluation of Four Onion Varieties in Montserrat*, Plymouth, CARDI, 1992.

Jones, R. M.: *Plant Resources of South-East Asia,* Wageningen, Pudoc Scientific Publishers, 1992.

Julia, F.: *Fruits of Warm Climates*, Miami, Julia F. Morton Publisher, 1987.

Kim, J. S.: *Sources of Resistance to Viruses of Pepper (Capsicum spp.): a Catalogue*, Taipei, AVRDC, 1994.

Kroll, R.: *Cut Flowers*, Wageningen, CTA, 1995.

Lorenzen, S.: *The Phylogenetic Systematics of Freeliving Nematodes*, London, The Ray Society, 1994.

Macself, A.J.: *Soils and Fertilizers*, Satish Serial Pub, Delhi, 2005.

Malins, A.: *Postharvest Handling of Pineapple and Mango*, Port of Spain, Trinidad and Tobago, 1992.

Mellor, J. W.: *The New Economics of Growth*, Ithaca, Cornell University Press, 1976.

Muneesh Kainth: *Chordate Embryology*, Dominant, Delhi, 2003.

Nobel, P. S.: *Physicochemical and Environmental Plant Physiology*, Academic Press, San Diego, 1999.

Oldham, P.: *Cost of Production of Major Tree Crops in Dominica*, Roseau, Ministry of Agriculture, 1991.

Parry, M.L.: *Climatic Change, Agriculture and Settlements*, Dawson Folkestone UK, 1978.

Pemberton, R. W.: *Predictable Risk to Native Plants in Weed Biological Control,* Oecologia, 2000.

Qystein, V. Sjaastad: *Physiology of Domestic Animals*, International Book Distributing Co., Delhi, 2005.

Ragone, D.: *Breadfruit: Artocarpus Altilis (Parkinson) Fosberg*, Rome, International Plant Genetic Resources Institute, 1997.

Sanchez, P. A.: *Properties and Management of Soils in the Tropics*, John Wiley and Sons, New York, 1976.

Steferud, Alfred: *Diseases of Fruits and Nuts*, Biotech Books, Delhi, 2005.

Tawde, A. B.: *Propagation and Rootstocks of Mango*, New Delhi, Malhotra, 1993.

Urton, Gary: *The Social Life of Numbers*, Austin, University of Texas Press, 1997.

Vanderplank, J.E.: *Plant Diseases: Epidemics and Control,* New York, NY, USA, Academic Press, 1963.

White, G.F.: *Natural Hazards: Local, National, Global*, Oxford University Press, New York, 1974.

Yoshida, T.: *Cultivation of Citrus Genetic Resources for Evaluation of Characteristics*, Tokyo, JICA, 1996.